T0229895

Interactive Web-Based Data Visualization with R, plotly, and shiny

Chapman & Hall/CRC
The R Series

Series Editors

John M. Chambers, Department of Statistics, Stanford University, California, USA
Torsten Hothorn, Division of Biostatistics, University of Zurich, Switzerland
Duncan Temple Lang, Department of Statistics, University of California, Davis, USA
Hadley Wickham, RStudio, Boston, Massachusetts, USA

Recently Published Titles

Analyzing Baseball Data with R, Second Edition
Max Marchi, Jim Albert, Benjamin S. Baumer

Spatio-Temporal Statistics with R
Christopher K. Wikle, Andrew Zammit-Mangion, and Noel Cressie

Statistical Computing with R, Second Edition
Maria L. Rizzo

Geocomputation with R
Robin Lovelace, Jakub Nowosad, Jannes Muenchow

Advanced R, Second Edition
Hadley Wickham

Dose Response Analysis Using R
Christian Ritz, Signe Marie Jensen, Daniel Gerhard, Jens Carl Streibig

Distributions for Modelling Location, Scale, and Shape
Using GAMLSS in R
Robert A. Rigby , Mikis D. Stasinopoulos, Gillian Z. Heller and Fernanda De Bastiani

Hands-On Machine Learning with R
Bradley Boehmke and Brandon Greenwell

Statistical Inference via Data Science
A ModernDive into R and the Tidyverse
Chester Ismay and Albert Y. Kim

Reproducible Research with R and RStudio, Third Edition
Christopher Gandrud

Interactive Web-Based Data Visualization with R, plotly, and shiny
Carson Sievert

For more information about this series, please visit: https://www.crcpress.com/
Chapman--HallCRC-The-R-Series/book-series/CRCTHERSER

Interactive Web-Based Data Visualization with R, plotly, and shiny

Carson Sievert

CRC Press
Taylor & Francis Group
Boca Raton London New York

CRC Press is an imprint of the
Taylor & Francis Group, an **informa** business

A CHAPMAN & HALL BOOK

CRC Press
Taylor & Francis Group
6000 Broken Sound Parkway NW, Suite 300
Boca Raton, FL 33487-2742

© 2020 by Taylor & Francis Group, LLC
CRC Press is an imprint of Taylor & Francis Group, an Informa business

No claim to original U.S. Government works

Printed on acid-free paper

International Standard Book Number-13: 978-1-138-33145-7 (Paperback)
International Standard Book Number-13: 978-1-138-33149-5 (Hardback)

Visit the Taylor & Francis Web site at
http://www.taylorandfrancis.com

and the CRC Press Web site at
http://www.crcpress.com

For the R community: Your kindness and generosity are truly inspiring.

Contents

1

Introduction

1.1   Why interactive web graphics *from R?*

As Wickham and Grolemund (2016) argue, the exploratory phase of a data science workflow (Figure 1.1) requires lots of iteration between data manipulation, visualization, and modeling. Achieving these tasks through a programming language like R offers the opportunity to scale and automate tasks, document and track them, and reliably reproduce their output. That power, however, typically comes at the cost of increasing the amount of cognitive load involved relative to a GUI-based system.[1] R packages like the **tidyverse** have been incredibly successful due to their ability to limit cognitive load without removing the benefits of performing analysis via code. Moreover, the **tidyverse**'s unifying principles of designing for humans, consistency, and composability make iteration within and between these stages seamless – an important but often overlooked challenge in exploratory data analysis (EDA) (Tidyverse team, 2018).

---

[1]For more on the benefits of using code over a GUI to perform data analysis, see Wickham (2018b).

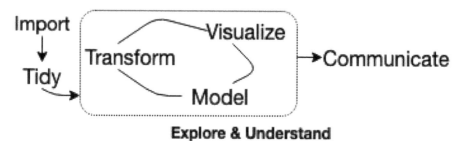

FIGURE 1.1: The stages of a data science workflow from Wickham and Grolemund (2016).

In fact, packages within the **tidyverse** such as **dplyr** (transformation) and **ggplot2** (visualization) are such productive tools that many analysts use *static* **ggplot2** graphics for EDA. Then, when it comes to communicating results, some analysts switch to another tool or language altogether (e.g., JavaScript) to generate interactive web graphics presenting their most important findings (Yau, 2016; Quealy, 2013). Unfortunately, this requires a heavy context switch that requires a totally different skillset and impedes productivity. Moreover, for the average analyst, the opportunity costs involved with becoming competent with the complex world of web technologies is simply not worth the required investment.

Even before the web, interactive graphics were shown to have great promise in aiding the exploration of high-dimensional data (Cook et al., 2007). The ASA maintains an incredible video library, http://stat-graphics.org/movies/, documenting the use of interactive statistical graphics for tasks that otherwise wouldn't have been easy or possible using numerical summaries and/or static graphics alone. Roughly speaking, these tasks tend to fall under three categories:

- Identifying structure that would otherwise go missing (Tukey and Fisherkeller, 1973).
- Diagnosing models and understanding algorithms (Wickham et al., 2015).

- Aiding the sense-making process by searching for information quickly without fully specified questions (Unwin and Hofmann, 1999).

Today, you can find and run some of these and similar Graphical User Interface (GUI) systems for creating interactive graphics: `DataDesk` `https://datadescription.com/`, `GGobi` `http://www.ggobi.org/`, `Mondrian` `http://www.theusrus.de/Mondrian/`, `JMP` `https://www.jmp.com`, `Tableau` `https://www.tableau.com/`. Although these GUI-based systems have nice properties, they don't gel with a code-based workflow: any tasks you complete through a GUI likely can't be replicated without human intervention. That means, if at any point, the data changes, and analysis outputs must be regenerated, you need to remember precisely how to reproduce the outcome, which isn't necessarily easy, trustworthy, or economical. Moreover, GUI-based systems are typically 'closed' systems that don't allow themselves to be easily customized, extended, or integrated with another system.

Programming interactive graphics allows you to leverage all the benefits of a code-based workflow while also helping with tasks that are difficult to accomplish with code alone. For an example, if you were to visualize engine displacement (`displ`) versus miles per gallon (`hwy`) using the `mpg` dataset, you might wonder: "what are these cars with an unusually high value of `hwy` given their `displ`?". Rather than trying to write code to query those observations, it would be easier and more intuitive to draw an outline around the points to query the data behind them.

```
library(ggplot2)
ggplot(mpg, aes(displ, hwy)) + geom_point()
```

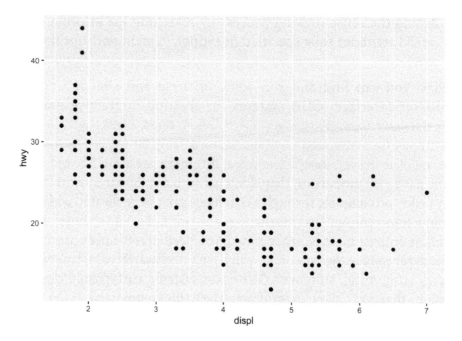

FIGURE 1.2: A scatterplot of engine displacement versus miles per gallon made with the **ggplot2** package.

Figure 1.3 demonstrates how we can transform Figure 1.2 into an interactive version that can be used to query and inspect points of interest. The framework that enables this kind of linked brushing is discussed in depth within Section 16.1, but the point here is that the added effort required to enable such functionality is relatively small. This is important, because although interactivity *can* augment exploration by allowing us to pursue follow-up questions, it's typically only *practical* when we can create and alter them quickly. That's because, in a true exploratory setting, you have to make lots of visualizations, and investigate lots of follow-up questions, before stumbling across something truly valuable.

```
library(plotly)
m <- highlight_key(mpg)
p <- ggplot(m, aes(displ, hwy)) + geom_point()
gg <- highlight(ggplotly(p), "plotly_selected")
crosstalk::bscols(gg, DT::datatable(m))
```

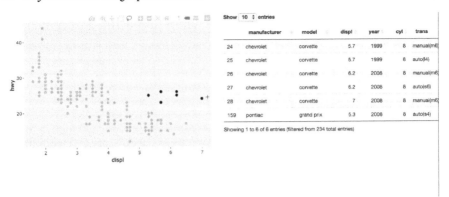

FIGURE 1.3: Linked brushing in a scatterplot to query more information about points of interest. By lasso selecting a region of unusual points, we learn that the corvette model has unusually high miles per gallon considering the engine size. For a video demonstration of the interactive, see `https://bit.ly/mpg-lasso`. For the interactive, see `https://plotly-r.com/interactives/mpg-lasso.html`

When a valuable insight surfaces, since the code behind Figure 1.3 generates HTML, the web-based graphic can be easily shared with collaborators through email and/or incorporated inside a larger automated report or website. Moreover, since these interactive graphics are based on the **htmlwidgets** framework, they work seamlessly inside larger **rmarkdown** documents, inside **shiny** apps, RStudio, Jupyter notebooks, the R prompt, and more. Being able to share interactive graphics with collaborators through these different mediums enhances the conversation – your colleagues can point out things you may not yet have considered and, in some cases, they can get immediate responses from the graphics themselves.

In the final stages of an analysis, when it comes time to publish your work to a general audience, rather than relying on the audience to interact with the graphics and discover insight for themselves, it's always a good idea to clearly highlight your findings. For example, from Figure 1.3, we've learned that most of these unusual points can be explained by a single feature of the data (`model == 'corvette'`). As shown in Figure 1.4, the `geom_mark_hull()` function from the **ggforce** package provides a helpful way to annotate those points with a hull. Moreover, as Chapter 12 demonstrates, it can also be helpful to add and/or edit annotations interactively when preparing a graphic for publication.

```
library(ggforce)
ggplot(mpg, aes(displ, hwy)) +
  geom_point() +
  geom_mark_hull(aes(filter = model == "corvette", label = model)) +
  labs(
    title = "Fuel economy from 1999 to 2008 for 38 car models",
    caption = "Source: https://fueleconomy.gov/",
    x = "Engine Displacement",
    y = "Miles Per Gallon"
  )
```

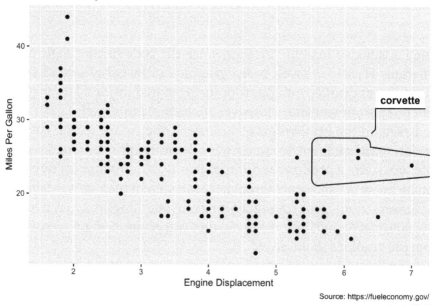

FIGURE 1.4: Using the **ggforce** package to annotate the corvettes in this dataset.

This simple example quickly shows how interactive web graphics can assist EDA (for another, slightly more in-depth example, see Section 2.3). Being able to program these graphics from ʀ allows one to combine their functionality within a world-class computing environment for data analysis and statistics. Programming interactive graphics may not be as intuitive as using a GUI-based system, but making the investment pays dividends in terms of workflow improvements: automation, scaling, provenance, and flexibility.

1.2   What you will learn

This book provides a foundation for learning how to make interactive web-based graphics for data analysis from R via **plotly**, without assuming any prior experience with web technologies. The goal is to provide the context you need to go beyond copying existing **plotly** examples to having a useful mental model of the underlying framework, its capabilities, and how it fits into the larger R ecosystem. By learning this mental model, you'll have a better understanding of how to create more sophisticated visualizations, fix common issues, improve performance, understand the limitations, and even contribute back to the project itself. You may already be familiar with existing **plotly** documentation (e.g., `https://plot.ly/r/`), which is essentially a language-agnostic how-to guide, but this book is meant to be a more holistic tutorial written by and for the R user.

This book also focuses primarily on features that are unique to the **plotly** R package (i.e., things that don't work the same for Python or JavaScript). This ranges from creation of a single graph using the `plot_ly()` special named arguments that make it easier to map data to visuals, as shown in Figure 1.5 (see more in Section 2.1). To its ability to link multiple data views purely client-side, as shown in Figure 1.6 (see more in Section 16.1). To advanced server-side linking with **shiny** to implement responsive and scalable crossfilters, as shown in Figure 1.7 (see more in Section 17.4.2).

```
plot_ly(diamonds, x = ~cut, color = ~clarity, colors = "Accent")
```

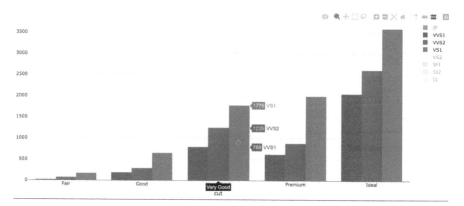

FIGURE 1.5: An example of what you'll learn: Figure 2.7. For a video demonstration of the interactive, see `https://bit.ly/intro-show-hide`. For the interactive, see `https://plotly-r.com/interactives/intro-show-hide.html`

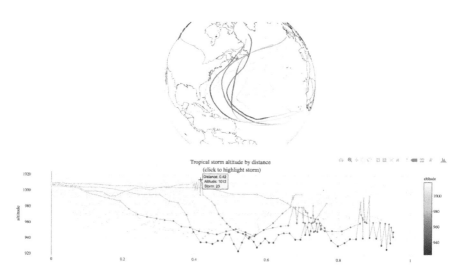

FIGURE 1.6: An example of what you'll learn: Figure 16.21. For a video demonstration of the interactive, see `https://bit.ly/storms-preview`. For the interactive, see `https://plotly-r.com/interactives/storms.html`

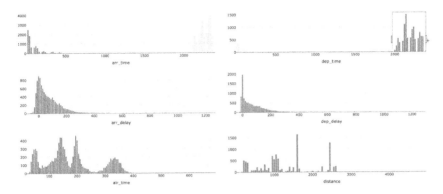

FIGURE 1.7: An example of what you'll learn: Figure 17.28. For a video demonstration of the interactive, see `https://bit.ly/shiny-crossfilter`. For the interactive, see `https://plotly-r.com/interactives/shiny-crossfilter.html`

By going through the code behind these examples, you'll see that many of them leverage other R packages in their implementation. To highlight a few of the R packages that you'll see:

- **dplyr** and **tidyr**
  – For transforming data into a form suitable for the visualization method.
- **ggplot2** and friends (e.g., **GGally**, **ggmosaic**, etc.)
  – For creating **plotly** visualizations that would be tedious to implement without `ggplotly()`.
- **sf**, **rnaturalearth**, **cartogram**
  – For obtaining and working with geo-spatial data structures in R.
- **stats**, **MASS**, **broom**, and **forecast**
  – For working with statistical models and summaries.
- **shiny**
  – For running R code in response to user input.
- **htmltools**, **htmlwidgets**
  – For combining multiple views and saving the result.

This book contains six parts and each part contains numerous chapters. A summary of each part is provided below.

1. *Creating views:* introduces the process of transforming data into graphics via **plotly**'s programmatic interface. It focuses mostly on `plot_ly()`, which can interface directly with the underlying plotly.js graphing library, but emphasis is put on features unique to the R package that make it easier to transform data into graphics. Another way to create graphs with **plotly** is to use the `ggplotly()` function to transform **ggplot2** graphs into **plotly** graphs. Section 2.3 discusses when and why `ggplotly()` might be desirable to `plot_ly()`. It's also worth mentioning that this part (nor the book as a whole) does not intend to cover every possible chart type and option available in **plotly** – it's more of a presentation of the most generally useful techniques with the greater R ecosystem in mind. For a more exhaustive gallery of examples of what **plotly** itself is capable of, see `https://plot.ly/r/`.

2. *Publishing views:* discusses various techniques for exporting (as well as embedding) **plotly** graphs to various file formats (e.g., HTML, SVG, PDF, PNG, etc.). Also, Chapter 12 demonstrates how one could leverage editable layout components HTML to touch up a graph, then export to a static file format of interest before publication. Indeed, this book was created using the techniques from this section.

3. *Combining multiple views:* demonstrates how to combine multiple data views into a single webpage (arranging) or graphic (animation). Most of these techniques are shown using **plotly** graphs, but techniques from Section 13.2 extend to any HTML content generated via **htmltools** (which includes **htmlwidgets**).

4. *Linking multiple views:* provides an overview of the two models for linking **plotly** graph(s) to other data views. The first model, covered in Section 16.1, outlines **plotly**'s support for linking views purely client-side, meaning the resulting graphs render in any web browser on any machine without requiring external software. The second model, covered in Chapter 17, demonstrates how to link **plotly** with other views via **shiny**, a reactive web application framework for R. Rel-

atively speaking, the second model grants the R user way more power and flexibility, but comes at the cost of requiring more computational infrastructure. That being said, RStudio provides accessible resources for deploying **shiny** apps `https://shiny.rstudio.com/articles/#deployment`.

5. *Custom behavior with JavaScript:* demonstrates various ways to customize **plotly** graphs by writing custom JavaScript to handle certain user events. This part of the book is designed to be approachable for R users that want to learn just enough JavaScript to **plotly** to do something it doesn't "natively" support.

6. *Various special topics:* offers a grab-bag of topics that address common questions, mostly related to the customization of **plotly** graphs in R.

You might already notice that this book often uses the term 'view' or 'data view', so here we take a moment to frame its use in a wider context. As Wills (2008) puts it: "a 'data view' is anything that gives the user a way of examining data so as to gain insight and understanding. A data view is usually thought of as a barchart, scatterplot, or other traditional statistical graphic, but we use the term more generally, including 'views' such as the results of a regression analysis, a neural net prediction, or a set of descriptive statistics". In this book, more often than not, the term 'view' typically refers to a **plotly** graph or other **htmlwidgets** (e.g., **DT**, **leaflet**, etc.). In particular, Section 16.1 is all about linking multiple **htmlwidgets** together through a graphical database querying framework. However, the term 'view' takes on a more general interpretation in Chapter 17 since the reactive programming framework that **shiny** provides allows us to have a more general conversation surrounding linked data views.

1.3   What you won't learn (much of)

1.3.1   Web technologies

Although this book is fundamentally about creating web graphics, it does not aim to teach you web technologies (e.g., HTML, SVG, CSS, JavaScript, etc.). It's true that mastering these technologies grants you the ability to build really impressive websites, but even expert web developers would say their skillset is much better suited for expository rather than exploratory visualization. That's because, most web programming tools are not well-suited for the exploratory phase of a data science workflow where iteration between data visualization, transformation, and modeling is a necessary task that often impedes hypothesis generation and sense-making. As a result, for most data analysts whose primary function is to derive insight from data, the opportunity costs involved with mastering web technologies is usually not worth the investment.

That being said, learning a little about web technologies can have a relatively large payoff with directed learning and instruction. In Chapter 18, you'll learn how to customize **plotly** graphs with JavaScript – even if you haven't seen JavaScript before, this chapter should be approachable, insightful, and provide you with some useful examples.

1.3.2   d3js

The JavaScript library D3 is a great tool for data visualization assuming you're familiar with web technologies and are primarily interested in expository (not exploratory) visualization. There are already lots of great resources for learning D3, including the numerous books by Murray (2013) and Murray (2017). It's worth noting, however, if you do know D3, you can easily leverage it from a webpage that is already a **plotly** graph, as demonstrated in Figure 22.1.

1.3.3   ggplot2

The book does contain some **ggplot2** code examples (which are then converted to **plotly** via `ggplotly()`), but it's not designed to teach you

ggplot2. For those looking to learn **ggplot2**, I recommend using the learning materials listed at `https://ggplot2.tidyverse.org`.

1.3.4 Graphical data analysis

How to perform data analysis via graphics (carefully, correctly, and creatively) is a large topic in itself. Although this book does have examples of graphical data analysis, it does not aim to provide a comprehensive foundation. For nice comprehensive resources on the topic, see Unwin (2015) and Cook and Swayne (2007).

1.3.5 Data visualization best practices

Encoding information in a graphic (concisely and effectively) is a large topic unto itself. Although this book does have some ramblings related to best practices in data visualization, it does not aim to provide a comprehensive foundation. For some approachable and fun resources on the topic, see Tufte (2001), Yau (2011), Healey (2018), and Wilke (2018).

1.4 Prerequisites

For those new to R and/or data visualization, R for Data Science[2] provides an excellent foundation for understanding the vast majority of concepts covered in this book (Wickham and Grolemund, 2016). In particular, if you have a solid grasp on Part I: Explore[3], Part II: Wrangle[4], and Part III: Program[5], you should be able to understand almost everything here. Although not explicitly covered, the book does make references to (and was creating using) **rmarkdown**, so if you're new to **rmarkdown**, I also recommend reading the R Markdown chapter[6].

---

[2]`https://r4ds.had.co.nz/`
[3]`https://r4ds.had.co.nz/explore-intro.html`
[4]`https://r4ds.had.co.nz/wrangle-intro.html`
[5]`https://r4ds.had.co.nz/program-intro.html`
[6]`https://r4ds.had.co.nz/r-markdown.html`

1.5   Run code examples

This book contains many code examples in an effort to teach the art and science behind creating interactive web-based graphics using **plotly**. To interact with the code results, you may either: (1) click on the static graphs hosted online at `https://plotly-r.com` and/or execute the code in a suitable computational environment. Most code examples assume you already have the **plotly** package loaded:

```
library(plotly)
```

If a particular code chunk doesn't work, you may need to load packages from previous examples in the chapter (some examples assume you're following the chapter in a linear fashion).

If you'd like to run examples on your local machine (instead of RStudio Cloud), you can install all the necessary R packages with:

```
if (!require(remotes)) install.packages("remotes")
remotes::install_github("cpsievert/plotly_book")
```

Visit `http://bit.ly/plotly-book-cloud` for a cloud-based instance of RStudio with all the required software to run the code examples in this book.

1.6   Getting help and learning more

As Wickham and Grolemund (2016) states, "This book is not an island; there is no single resource that will allow you to master R [or **plotly**]. As you start to apply the techniques described in this book to your own data, you will soon find questions that I do not answer. This section describes a few tips on how to get help, and to help you keep learning."

These tips[7] on how to get help (e.g., Google, StackOverflow, Twitter, etc.) also apply to getting help with **plotly**. RStudio's community[8] is another great place to ask broader questions about all things R and **plotly**. It's worth mentioning that the R community is incredibly welcoming, compassionate, and generous; especially if you can demonstrate that you've done your research and/or provide a minimally reproducible example of your problem[9].

1.7 Acknowledgments

This book wouldn't be possible without the generous assistance and mentorship of many people:

- Heike Hofmann and Di Cook for their mentorship and many helpful conversations about interactive graphics.
- Toby Dylan Hocking for many helpful conversations, his mentorship in the R packages **animint** and **plotly**, and laying the original foundation behind ggplotly().
- Joe Cheng for many helpful conversations and inspiring Section 16.1.
- Étienne Tétreault-Pinard, Alex Johnson, and the other plotly.js core developers for responding to my feature requests and bug reports.
- Yihui Xie for his work on **knitr**, **rmarkdown**, **bookdown**, bookdown-crc[10], and responding to my feature requests.
- Anthony Unwin for helpful feedback, suggestions, and for inspiring Figure 16.13.
- Hadley Wickham and the **ggplot2** team for maintaining **ggplot2**.
- Hadley Wickham and Garret Grolemund for writing *R for Data Science* and allowing me to model this introduction after their introduction.
- Kent Russell for contributions to **plotly**, **htmlwidgets**, and **reactR**.
- Adam Loy for inspiring Figure 14.5.

---

[7]https://r4ds.had.co.nz/introduction.html#getting-help-and-learning-more
[8]https://community.rstudio.com/tags/plotly
[9]https://www.tidyverse.org/help/
[10]https://github.com/yihui/bookdown-crc

- Many other R community members who contributed to the **plotly** package and provided feedback and corrections for this book.

1.8   Colophon

An online version of this book is available at `https://plotly-r.com`. It will continue to evolve in between reprints of the physical book. The source of the book is available at `https://github.com/cpsievert/plotly_book`. The book is powered by `https://bookdown.org` which makes it easy to turn R markdown files into HTML, PDF, and EPUB.

This book was built with the following computing environment:

```
devtools::session_info("plotly")
#> - Session info ------------------------------------------
#>  setting  value
#>  version  R version 3.6.1 (2019-07-05)
#>  os       macOS Mojave 10.14.5
#>  system   x86_64, darwin15.6.0
#>  ui       X11
#>  language (EN)
#>  collate  en_US.UTF-8
#>  ctype    en_US.UTF-8
#>  tz       America/Chicago
#>  date     2019-10-07
#>
#> - Packages ----------------------------------------------
#>  package     * version   date         lib
#>  askpass       1.1       2019-01-13 [1]
#>  assertthat    0.2.1     2019-03-21 [1]
#>  backports     1.1.5     2019-10-02 [1]
#>  base64enc     0.1-3     2015-07-28 [1]
#>  BH            1.69.0-1  2019-01-07 [1]
#>  cli           1.1.0     2019-03-19 [1]
#>  colorspace    1.4-1     2019-03-18 [1]
```

```
#>   crayon         1.3.4        2017-09-16 [1]
#>   crosstalk      1.0.0        2016-12-21 [1]
#>   curl           4.2          2019-09-24 [1]
#>   data.table     1.12.4       2019-10-03 [1]
#>   digest         0.6.21       2019-09-20 [1]
#>   dplyr          0.8.3        2019-07-04 [1]
#>   ellipsis       0.3.0        2019-09-20 [1]
#>   fansi          0.4.0        2018-10-05 [1]
#>   ggplot2      * 3.2.1.9000   2019-10-07 [1]
#>   glue           1.3.1        2019-03-12 [1]
#>   gtable         0.3.0        2019-03-25 [1]
#>   hexbin         1.27.3       2019-05-14 [1]
#>   htmltools      0.3.6.9004   2019-10-07 [1]
#>   htmlwidgets    1.5.1        2019-10-07 [1]
#>   httpuv         1.5.2        2019-09-11 [1]
#>   httr           1.4.1        2019-08-05 [1]
#>   jsonlite       1.6          2018-12-07 [1]
#>   labeling       0.3          2014-08-23 [1]
#>   later          1.0.0        2019-09-17 [1]
#>   lattice        0.20-38      2018-11-04 [1]
#>   lazyeval       0.2.2        2019-03-15 [1]
#>   lifecycle      0.1.0        2019-08-01 [1]
#>   magrittr       1.5          2014-11-22 [1]
#>   MASS           7.3-51.4     2019-03-31 [1]
#>   Matrix         1.2-17       2019-03-22 [1]
#>   mgcv           1.8-29       2019-09-20 [1]
#>   mime           0.7          2019-06-11 [1]
#>   munsell        0.5.0        2018-06-12 [1]
#>   nlme           3.1-141      2019-08-01 [1]
#>   openssl        1.4.1        2019-07-18 [1]
#>   pillar         1.4.2        2019-06-29 [1]
#>   pkgconfig      2.0.3        2019-09-22 [1]
#>   plogr          0.2.0        2018-03-25 [1]
#>   plotly       * 4.9.0.9000   2019-10-07 [1]
#>   plyr           1.8.4        2016-06-08 [1]
#>   promises       1.1.0        2019-09-25 [1]
#>   purrr          0.3.2.9000   2019-09-30 [1]
```

```
#>  R6              2.4.0        2019-02-14 [1]
#>  RColorBrewer    1.1-2        2014-12-07 [1]
#>  Rcpp            1.0.2        2019-07-25 [1]
#>  reshape2        1.4.3        2017-12-11 [1]
#>  rlang           0.4.0.9003 2019-10-07 [1]
#>  scales          1.0.0.9000 2019-07-19 [1]
#>  shiny           1.3.2        2019-10-07 [1]
#>  sourcetools     0.1.7        2018-04-25 [1]
#>  stringi         1.4.3        2019-03-12 [1]
#>  stringr         1.4.0        2019-02-10 [1]
#>  sys             3.3          2019-08-21 [1]
#>  tibble          2.1.3        2019-06-06 [1]
#>  tidyr           1.0.0        2019-09-11 [1]
#>  tidyselect      0.2.5        2018-10-11 [1]
#>  utf8            1.1.4        2018-05-24 [1]
#>  vctrs           0.2.0.9003 2019-10-01 [1]
#>  viridisLite     0.3.0        2018-02-01 [1]
#>  withr           2.1.2        2018-03-15 [1]
#>  xtable          1.8-4        2019-04-21 [1]
#>  yaml            2.2.0        2018-07-25 [1]
#>  zeallot         0.1.0        2018-01-28 [1]
#>  source
#>  CRAN (R 3.6.0)
#>  CRAN (R 3.6.0)
#>  CRAN (R 3.6.0)
#>  CRAN (R 3.6.0)
#>  CRAN (R 3.6.0)
#>  CRAN (R 3.6.0)
#>  CRAN (R 3.6.0)
#>  CRAN (R 3.6.0)
#>  CRAN (R 3.6.1)
#>  CRAN (R 3.6.1)
#>  CRAN (R 3.6.1)
#>  CRAN (R 3.6.0)
#>  CRAN (R 3.6.1)
#>  CRAN (R 3.6.0)
```

```
#>  local
#>  CRAN (R 3.6.0)
#>  CRAN (R 3.6.0)
#>  CRAN (R 3.6.0)
#>  local
#>  local
#>  CRAN (R 3.6.1)
#>  CRAN (R 3.6.1)
#>  CRAN (R 3.6.0)
#>  CRAN (R 3.6.0)
#>  Github (r-lib/later@0364de9)
#>  CRAN (R 3.6.1)
#>  CRAN (R 3.6.0)
#>  CRAN (R 3.6.0)
#>  CRAN (R 3.6.0)
#>  CRAN (R 3.6.1)
#>  CRAN (R 3.6.1)
#>  CRAN (R 3.6.0)
#>  CRAN (R 3.6.0)
#>  CRAN (R 3.6.0)
#>  CRAN (R 3.6.1)
#>  CRAN (R 3.6.0)
#>  CRAN (R 3.6.1)
#>  CRAN (R 3.6.0)
#>  local
#>  CRAN (R 3.6.0)
#>  Github (rstudio/promises@39faf86)
#>  Github (tidyverse/purrr@9edf0ca)
#>  CRAN (R 3.6.0)
#>  CRAN (R 3.6.0)
#>  CRAN (R 3.6.1)
#>  CRAN (R 3.6.0)
#>  Github (r-lib/rlang@09fda4a)
#>  Github (r-lib/scales@7f6f4a5)
#>  local
#>  CRAN (R 3.6.0)
```

```
#>   CRAN (R 3.6.0)
#>   CRAN (R 3.6.0)
#>   CRAN (R 3.6.0)
#>   CRAN (R 3.6.0)
#>   CRAN (R 3.6.0)
#>   CRAN (R 3.6.0)
#>   CRAN (R 3.6.0)
#>   Github (r-lib/vctrs@bc20422)
#>   CRAN (R 3.6.0)
#>   CRAN (R 3.6.0)
#>   CRAN (R 3.6.0)
#>   CRAN (R 3.6.0)
#>
#> [1] /Library/Frameworks/R.framework/Versions/3.6/Resources/library
```

Part I

Creating views

2

Overview

This part of the book teaches you how to leverage the **plotly** R package to create a variety of interactive graphics. There are two main ways to creating a **plotly** object: either by transforming a **ggplot2** object (via `ggplotly()`) into a **plotly** object or by directly initializing a **plotly** object with `plot_ly()`/`plot_geo()`/`plot_mapbox()`. Both approaches have somewhat complementary strengths and weaknesses, so it can pay off to learn both approaches. Moreover, both approaches are an implementation of the Grammar of Graphics and both are powered by the JavaScript graphing library plotly.js, so many of the same concepts and tools that you learn for one interface can be reused in the other.

The subsequent chapters within this 'Creating views' part dive into specific examples and use cases, but this introductory chapter outlines some over-arching concepts related to **plotly** in general. It also provides definitions for terminology used throughout the book and introduces some concepts useful for understanding the infrastructure behind any **plotly** object. Most of these details aren't necessarily required to get started with **plotly**, but it will inevitably help you get 'un-stuck', write better code, and do more advanced things with **plotly**.

2.1   Intro to `plot_ly()`

Any graph made with the **plotly** R package is powered by the JavaScript library plotly.js[1]. The `plot_ly()` function provides a 'direct' interface to plotly.js with some additional abstractions to help reduce typing. These abstractions, inspired by the Grammar of Graphics and **ggplot2**,

---

[1] https://github.com/plotly/plotly.js

make it much faster to iterate from one graphic to another, making it easier to discover interesting features in the data (Wilkinson, 2005; Wickham, 2009). To demonstrate, we'll use `plot_ly()` to explore the `diamonds` dataset from **ggplot2** and learn a bit how **plotly** and plotly.js work along the way.

```r
# load the plotly R package
library(plotly)
```

```r
# load the diamonds dataset from the ggplot2 package
data(diamonds, package = "ggplot2")
diamonds
#> # A tibble: 53,940 x 10
#>    carat cut    color clarity depth table price     x
#>    <dbl> <ord>  <ord> <ord>   <dbl> <dbl> <int> <dbl>
#> 1  0.23  Ideal  E     SI2     61.5     55   326  3.95
#> 2  0.21  Prem~  E     SI1     59.8     61   326  3.89
#> 3  0.23  Good   E     VS1     56.9     65   327  4.05
#> 4  0.290 Prem~  I     VS2     62.4     58   334   4.2
#> 5  0.31  Good   J     SI2     63.3     58   335  4.34
#> 6  0.24  Very~  J     VVS2    62.8     57   336  3.94
#> # ... with 5.393e+04 more rows, and 2 more variables:
#> #   y <dbl>, z <dbl>
```

If we assign variable names (e.g., `cut`, `clarity`, etc.) to visual properties (e.g., `x`, `y`, `color`, etc.) within `plot_ly()`, as done in Figure 2.1, it tries to find a sensible geometric representation of that information for us. Shortly we'll cover how to specify these geometric representations (as well as other visual encodings) to create different kinds of charts.

```r
# create three visualizations of the diamonds dataset
plot_ly(diamonds, x = ~cut)
plot_ly(diamonds, x = ~cut, y = ~clarity)
plot_ly(diamonds, x = ~cut, color = ~clarity, colors = "Accent")
```

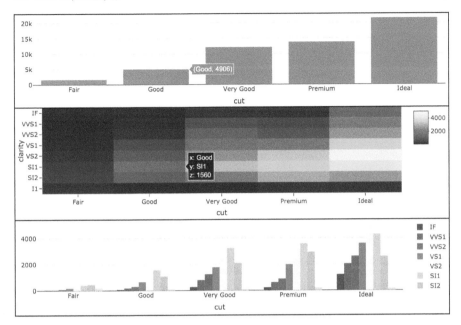

FIGURE 2.1: Three examples of visualizing categorical data with `plot_ly()`: (top) mapping `cut` to `x` yields a bar chart, (middle) mapping `cut` & `clarity` to `x` & `y` yields a heatmap, and (bottom) mapping `cut` & `clarity` to `x` & `color` yields a dodged bar chart.

The `plot_ly()` function has numerous arguments that are unique to the R package (e.g., `color`, `stroke`, `span`, `symbol`, `linetype`, etc.) and make it easier to encode data variables (e.g., diamond clarity) as visual properties (e.g., color). By default, these arguments map values of a data variable to a visual range defined by the plural form of the argument. For example, in the bottom panel of 2.1, `color` is used to map each level of diamond clarity to a different color, then `colors` is used to specify the range of colors (which, in this case, the `"Accent"` color palette from the **RColorBrewer** package, but one can also supply custom color codes or a color palette function like `colorRamp()`). Figure 2.2 provides a visual diagram of how this particular mapping works, but the same sort of idea can be applied to other visual properties like size, shape, linetype, etc.

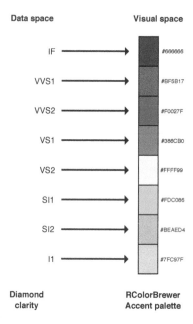

FIGURE 2.2: Mapping data values to a visual color range.

Since these arguments map data values to a visual range by default, you will obtain unexpected results if you try to specify the visual range directly, as in the top portion of Figure 2.3. If you want to specify the visual range directly, use the I() function to declare this value to be taken 'AsIs', as in the bottom portion of Figure 2.3. Throughout this book, you'll see lots of examples that leverage these arguments, especially in Chapter 3. Another good resource to learn more about these arguments (especially their defaults) is the R documentation page available by entering help(plot_ly) in your R console.

```
# doesn't produce black bars
plot_ly(diamonds, x = ~cut, color = "black")
# produces red bars with black outline
plot_ly(
  diamonds,
  x = ~cut,
  color = I("red"),
```

```
  stroke = I("black"),
  span = I(2)
)
```

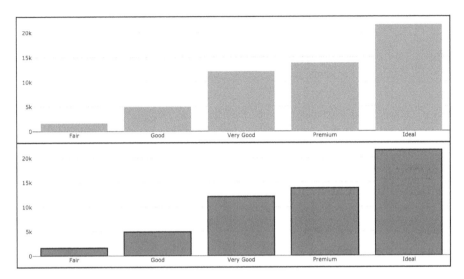

FIGURE 2.3: Using `I()` to supply visual properties directly instead of mapping values to a visual range. In the top portion of this figure, the value `'black'` is being mapped to a visual range spanned by `colors` (which, for discrete data, defaults to `'Set2'`).

The **plotly** package takes a purely functional approach to a layered grammar of graphics (Wickham, 2010).[2] The purely functional part means, (almost) every function anticipates a **plotly** object as input to its first argument and returns a modified version of that **plotly** object. Furthermore, that modification is completely determined by the input values to the function (i.e., it doesn't rely on any side effects, unlike, for example, base R graphics). For a quick example, the `layout()` function anticipates a **plotly** object in its first argument and its other arguments add and/or modify various layout components of that object (e.g., the title):

---

[2]If you aren't already familiar with the grammar of graphics or **ggplot2**, we recommend reading the Data Visualization chapter from the *R for Data Science* book. `https://r4ds.had.co.nz/data-visualisation.html`

```
layout(
  plot_ly(diamonds, x = ~cut),
  title = "My beatiful histogram"
)
```

For more complex plots that modify a **plotly** graph many times over, code written in this way can become cumbersome to read. In particular, we have to search for the innermost part of the R expression, then work outwards towards the end result. The %>% operator from the **magrittr** package allows us to rearrange this code so that we can read the sequence of modifications from left-to-right rather than inside-out (Bache and Wickham, 2014). The %>% operator enables this by placing the object on the left-hand side of the %>% into the first argument of the function of the right-hand side.

```
diamonds %>%
  plot_ly(x = ~cut) %>%
  layout(title = "My beatiful histogram")
```

In addition to layout() for adding/modifying part(s) of the graph's layout, there are also a family of add_*() functions (e.g., add_histogram(), add_lines(), etc.) that define how to render data into geometric objects. Borrowing terminology from the layered grammar of graphics, these functions add a graphical layer to a plot. A *layer* can be thought of as a group of graphical elements that can be sufficiently described using only 5 components: data, aesthetic mappings (e.g., assigning clarity to color), a geometric representation (e.g., rectangles, circles, etc.), statistical transformations (e.g., sum, mean, etc.), and positional adjustments (e.g., dodge, stack, etc.). If you're paying attention, you'll notice that in the examples thus far, we have not specified a layer! The layer has been added for us automatically by plot_ly(). To be explicit about what plot_ly(diamonds, x = ~cut) generates, we should add a add_histogram() layer:

```
diamonds %>%
  plot_ly() %>%
  add_histogram(x = ~cut)
```

As you'll learn more about in Chapter 5, **plotly** has both `add_histogram()` and `add_bars()`. The difference is that `add_histogram()` performs *statistics* (i.e., a binning algorithm) dynamically in the web browser, whereas `add_bars()` requires the bar heights to be pre-specified. That means, to replicate the last example with `add_bars()`, the number of observations must be computed ahead of time.

```
diamonds %>%
  dplyr::count(cut) %>%
  plot_ly() %>%
  add_bars(x = ~cut, y = ~n)
```

There are numerous other `add_*()` functions that calculate statistics in the browser (e.g., `add_histogram2d()`, `add_contour()`, `add_boxplot()`, etc.), but most other functions aren't considered statistical. Making the distinction might not seem useful now, but they have their own respective trade-offs when it comes to speed and interactivity. Generally speaking, non-statistical layers will be faster and more responsive at runtime (since they require less computational work), whereas the statistical layers allow for more flexibility when it comes to client-side interactivity, as covered in Chapter 16. Practically speaking, the difference in performance is often negligible. The more common bottleneck occurs when attempting to render lots of graphical elements at a time (e.g., a scatterplot with a million points). In those scenarios, you likely want to render your plot in Canvas rather than SVG (the default) via `toWebGL()`. For more information on improving performance, see Chapter 24.

In many scenarios, it can be useful to combine multiple graphical layers into a single plot. In this case, it becomes useful to know a few things about `plot_ly()`:

- Arguments specified in `plot_ly()` are *global*, meaning that any downstream `add_*()` functions inherit these arguments (unless `inherit = FALSE`).
- Data manipulation verbs from the **dplyr** package may be used to transform the `data` underlying a **plotly** object.[3]

Using these two properties of `plot_ly()`, Figure 2.4 demonstrates how we could leverage these properties of `plot_ly()` to do the following:

1.   *Globally* assign `cut` to x.
2.   Add a histogram layer (inherits the x from `plot_ly()`).
3.   Use **dplyr** verbs to modify the `data` underlying the **plotly** object. Here we just count the number of diamonds in each `cut` category.
4.   Add a layer of text using the summarized counts. Note that the global x mapping, as well as the other mappings local to this text layer (`text` and `y`), reflects data values from step 3.

```
library(dplyr)

diamonds %>%
  plot_ly(x = ~cut) %>%
  add_histogram() %>%
  group_by(cut) %>%
  summarise(n = n()) %>%
  add_text(
    text = ~scales::comma(n), y = ~n,
    textposition = "top middle",
    cliponaxis = FALSE
  )
```

---

[3]Technically speaking, these **dplyr** verbs are S3 generic functions that have a **plotly** method. In nearly every case, that method simply queries the data underlying the **plotly** object, applies the **dplyr** function, then adds the transformed data back into the resulting **plotly** object.

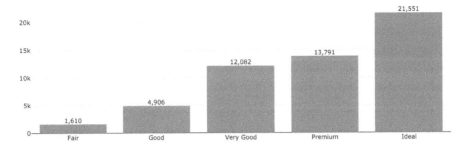

FIGURE 2.4: Using `add_histogram()`, `add_text()`, and **dplyr** verbs to compose a plot that leverages a raw form of the data (e.g., histogram) as well as a summarized version (e.g., text labels).

Before using multiple `add_*()` in a single plot, make sure that you actually want to show those layers of information on the same set of axes. If it makes sense to display the information on the same axes, consider making multiple **plotly** objects and combining them into a grid-like layout using `subplot()`, as described in Chapter 13. Also, when using **dplyr** verbs to modify the `data` underlying the **plotly** object, you can use the `plotly_data()` function to obtain the data at any point in time, which is primarily useful for debugging purposes (i.e., inspecting the data of a particular graphical layer).

```
diamonds %>%
  plot_ly(x = ~cut) %>%
  add_histogram() %>%
  group_by(cut) %>%
  summarise(n = n()) %>%
  plotly_data()
#> # A tibble: 5 x 2
#>   cut            n
#>   <ord>      <int>
#> 1 Fair        1610
#> 2 Good        4906
#> 3 Very Good  12082
#> 4 Premium    13791
#> 5 Ideal      21551
```

This introduction to `plot_ly()` has mainly focused on concepts unique to the R package **plotly** that are generally useful for creating most kinds of data views. The next section outlines how **plotly** generates plotly.js figures and how to inspect the underlying data structure that plotly.js uses to render the graph. Not only is this information useful for debugging, but it's also a nice way to learn how to work with plotly.js directly, which you may need to improve performance in **shiny** apps (Section 17.3.1) and/or for adding custom behavior with JavaScript (Chapter 18).

2.2   Intro to plotly.js

To recreate the plots in Figure 2.1 using plotly.js *directly*, it would take significantly more code and knowledge of plotly.js. That being said, learning how **plotly** generates the underlying plotly.js figure is a useful introduction to plotly.js itself, and knowledge of plotly.js becomes useful when you need more flexible control over **plotly**. As Figure 2.5 illustrates, when you print any **plotly** object, the `plotly_build()` function is applied to that object, and that generates an R list which adheres to a syntax that plotly.js understands. This syntax is a JavaScript Object Notation (JSON) specification that plotly.js uses to represent, serialize, and render web graphics. A lot of documentation you'll find online about plotly (e.g., the online figure reference[4]) implicitly refers to this JSON specification, so it can be helpful to know how to "work backwards" from that documentation (i.e., translate JSON into to R code). If you'd like to learn details about mapping between R and JSON, Chapter 19 provides an introduction aimed at R programmers, and Ooms (2014) provides a cohesive overview of the **jsonlite** package, which is what **plotly** uses to map between R and JSON.

---

[4]`https://plot.ly/r/reference/`

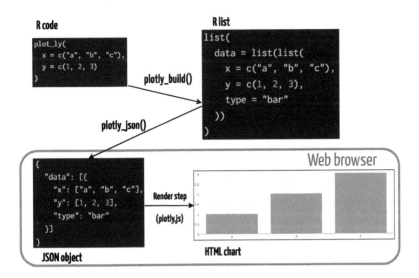

FIGURE 2.5: A diagram of what happens when you print a **plotly** graph.

For illustration purposes, Figure 2.5 shows how this workflow applies to a simple bar graph (with values directly supplied instead of a data column name reference like Figure 2.1), but the same concept applies for any graph created via **plotly**. As the diagram suggests, both the `plotly_build()` and `plotly_json()` functions can be used to inspect the underlying data structure on both the R and JSON side of things. For example, Figure 2.6 shows the `data` portion of the JSON created for the last graph in Figure 2.6.

```
p <- plot_ly(diamonds, x = ~cut, color = ~clarity, colors = "Accent")
plotly_json(p)
```

```
▼ data [8]
   ▶ 0   {9}
   ▼ 1   {9}
      ▶ x     [3655]
         type : histogram
         name : VVS1
      ▼ marker {2}
            color : ■ rgba(191,91,23,1)
```

FIGURE 2.6: A portion of the JSON data behind the bottom plot of Figure 2.1. This dodged bar chart has eight layers of data (i.e., eight traces), one for each level of clarity.

In plotly.js terminology, a *figure* has two key components: data (aka, traces) and a layout. A *trace* defines a mapping from data and visuals.[5] Every trace has a *type* (e.g., histogram, pie, scatter, etc.) and the trace type determines what other attributes (i.e., visual and/or interactive properties, like x, hoverinfo, name) are available to control the trace mapping. That is, not every trace attribute is available to every trace type, but many attributes (e.g., the name of the trace) are available in every trace type and serve a similar purpose. From Figure 2.6, we can see that it takes multiple traces to generate the dodged bar chart, but instead of clicking through JSON viewer, sometimes it's easier to use plotly_build() and compute on the plotly.js figure definition to verify certain things exist. Since **plotly** uses the **htmlwidgets** standard[6], the actual plotly.js figure definition appears under a list element named x (Vaidyanathan et al., 2016).

---

[5]A trace is similar in concept to a layer (as defined in Section 2.1), but it's not quite the same. In many cases, like the bottom panel of Figure 2.1, it makes sense to implement a single layer as multiple traces. This is due to the design of plotly.js and how traces are tied to legends and hover behavior.

[6]The **htmlwidgets** package provides a foundation for other packages to implement R bindings to JavaScript libraries so that those bindings work in various contexts (e.g., the R console, RStudio, inside **rmarkdown** documents, **shiny** apps, etc.). For more info and examples, see the website http://www.htmlwidgets.org.

```
# use plotly_build() to get at the plotly.js definition
# behind *any* plotly object
b <- plotly_build(p)


# Confirm there 8 traces
length(b$x$data)
#> [1] 8


# Extract the `name` of each trace. plotly.js uses `name` to
# populate legend entries and tooltips
purrr::map_chr(b$x$data, "name")
#> [1] "IF" "VVS1" "VVS2" "VS1" "VS2" "SI1" "SI2" "I1"


# Every trace has a type of histogram
unique(purrr::map_chr(b$x$data, "type"))
#> [1] "histogram"
```

Here we've learned that **plotly** creates 8 histogram traces to generate
the dodged bar chart: one trace for each level of clarity.[7] Why one trace
per category? As illustrated in Figure 2.7, there are two main reasons:
to populate a tooltip and legend entry for each level of clarity level.

---

[7]Although the x-axis is discrete, plotly.js still considers this a histogram because
it generates counts in the browser. Learn more about the difference between histograms and bar charts in Chapter 5.

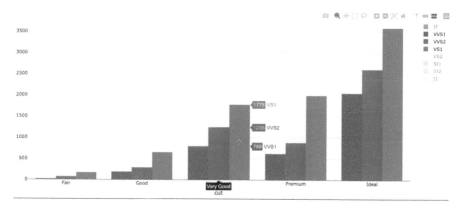

FIGURE 2.7: Leveraging two interactive features that require one trace per level of `clarity`: (1) Using 'Compare data on hover' mode to get counts for every level of `clarity` for a given level of `cut`, and (2) using the ability to hide/show clarity levels via their legend entries. For a video demonstration of the interactive, see `https://bit.ly/intro-show-hide-preview`. For the interactive, see `https://plotly-r.com/interactives/intro-show-hide.html`

If we investigated further, we'd notice that `color` and `colors` are not officially part of the plotly.js figure definition; the `plotly_build()` function has effectively transformed that information into a sensible plotly.js figure definition (e.g., `marker.color` contains the actual bar color codes). In fact, the `color` argument in `plot_ly()` is just one example of an abstraction the R package has built on top of plotly.js to make it easier to map data values to visual attributes, and many of these are covered in Chapter 3.

2.3   Intro to `ggplotly()`

The `ggplotly()` function from the **plotly** package has the ability to translate **ggplot2** to **plotly**. This functionality can be really helpful for quickly adding interactivity to your existing **ggplot2** workflow.[8] Moreover, even if you know plot_ly() and plotly.js well, ggplotly() can still be desirable for creating visualizations that aren't necessarily straightforward to achieve without it. To demonstrate, let's explore the relationship between price and other variables from the well-known diamonds dataset.

Hexagonal binning (i.e., geom_hex()) is useful way to visualize a 2D density[9], like the relationship between price and carat as shown in Figure 2.8. From Figure 2.8, we can see there is a strong positive linear relationship between the *log* of carat and price. It also shows that for many, the carat is only rounded to a particular number (indicated by the light blue bands) and no diamonds are priced around $1500. Making this plot interactive makes it easier to decode the hexagonal colors into the counts that they represent.

```
p <- ggplot(diamonds, aes(x = log(carat), y = log(price))) +
  geom_hex(bins = 100)
ggplotly(p)
```

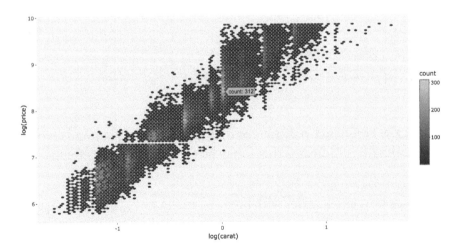

FIGURE 2.8: A hexbin plot of diamond carat versus price.

---

[8]This section is not meant to teach you **ggplot2**, but rather to help point out when and why it might be preferable to plot_ly(). If you're new to **ggplot2** and would like to learn it, see Section 1.3.3.
[9]See Section 7 for approaches using plot_ly()

I often use `ggplotly()` over `plot_ly()` to leverage **ggplot2**'s consistent and expressive interface for exploring statistical summaries across groups. For example, by including a discrete `color` variable (e.g., `cut`) with `geom_freqpoly()`, you get a frequency polygon for each level of that variable. This ability to quickly generate visual encodings of statistical summaries across an arbitrary number of groups works for basically any geom (e.g., `geom_boxplot()`, `geom_histogram()`, `geom_density()`, etc.) and is a key feature of **ggplot2**.

```
p <- ggplot(diamonds, aes(x = log(price), color = clarity)) +
    geom_freqpoly()
ggplotly(p)
```

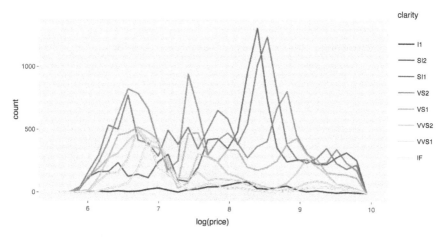

FIGURE 2.9: Frequency polygons of diamond price by diamond clarity. This visualization indicates there may be significant main effects.

Now, to see how `price` varies with both `cut` and `clarity`, we could repeat this same visualization for each level of `cut`. This is where **ggplot2**'s `facet_wrap()` comes in handy. Moreover, to facilitate comparisons, we can have `geom_freqpoly()` display relative rather than absolute frequencies. By making this plot interactive, we can more easily compare particular levels of clarity (as shown in Figure 2.10) by leveraging the legend filtering capabilities.

```
p <- ggplot(diamonds, aes(x = log(price), color = clarity)) +
    geom_freqpoly(stat = "density") +
    facet_wrap(~cut)
ggplotly(p)
```

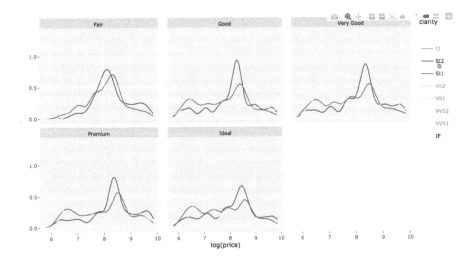

FIGURE 2.10: Diamond price by clarity and cut. For a video demonstration of the interactive, see `https://bit.ly/freqpoly-facet`. For the interactive, see `https://plotly-r.com/interactives/freqpoly-facet.html`

In addition to supporting most of the 'core' **ggplot2** API, `ggplotly()` can automatically convert any **ggplot2** extension packages that return a 'standard' **ggplot2** object. By standard, I mean that the object is comprised of 'core' **ggplot2** data structures and not the result of custom geoms.[10] Some great examples of R packages that extend **ggplot2** using core data structures are **ggforce**, **naniar**, and **GGally** (Pedersen, 2019; Tierney et al., 2018; Schloerke et al., 2016).

Figure 2.11 demonstrates another way of visualizing the same information found in Figure 2.10 using `geom_sina()` from the **ggforce** pack-

---

[10]As discussed in Chapter 34, `ggplotly()` can actually convert custom geoms as well, but each one requires a custom hook, and many custom geoms are not yet supported.

age (instead of `geom_freqpoly()`). This visualization jitters the raw data within the density for each group allowing us not only to see where the majority observations fall within a group, but also across all groups. By making this layer interactive, we can query individual points for more information and zoom into interesting regions. The second layer of Figure 2.11 uses **ggplot2**'s `stat_summary()` to overlay a 95% confidence interval estimated via a Bootstrap algorithm via the **Hmisc** package (Harrell Jr et al., 2019).

```
p <- ggplot(diamonds, aes(x=clarity, y=log(price), color=clarity)) +
    ggforce::geom_sina(alpha = 0.1) +
    stat_summary(fun.data = "mean_cl_boot", color = "black") +
    facet_wrap(~cut)

# WebGL is a lot more efficient at rendering lots of points
toWebGL(ggplotly(p))
```

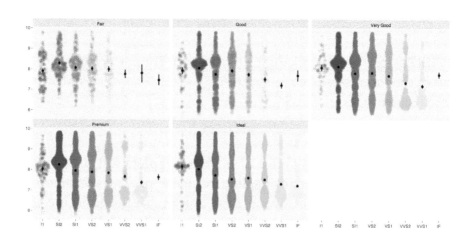

FIGURE 2.11: A sina plot of diamond price by clarity and cut.

As noted by Wickham and Grolemund (2016), it's surprising that the diamond price would decline with an increase of diamond clarity. As it turns out, if we account for the carat of the diamond, then we see that better diamond clarity does indeed lead to a higher diamond price, as shown in Figure 2.12. Seeing such a strong pattern in the residuals of simple linear model of carat vs. price indicates that our model could be greatly improved by adding `clarity` as a predictor of `price`.

```r
m <- lm(log(price) ~ log(carat), data = diamonds)
diamonds <- modelr::add_residuals(diamonds, m)
p <- ggplot(diamonds, aes(x = clarity, y = resid, color = clarity)) +
    ggforce::geom_sina(alpha = 0.1) +
    stat_summary(fun.data = "mean_cl_boot", color = "black") +
    facet_wrap(~cut)
toWebGL(ggplotly(p))
```

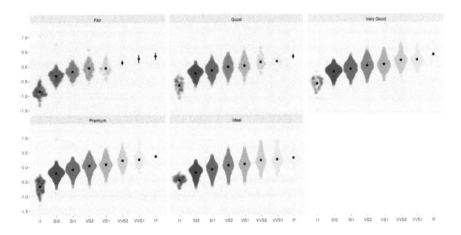

FIGURE 2.12: A sina plot of diamond price by clarity and cut, after accounting for carat.

As discussed in Section 16.4.7, the **GGally** package provides a convenient interface for making similar types of model diagnostic visualizations via the `ggnostic()` function. It also provides a convenience function for visualizing the coefficient estimates and their standard errors via the `ggcoef()` function. Figure 2.13 shows how injecting interactivity into this plot allows us to query exact values and zoom in on the most interesting regions.

```
library(GGally)
m <- lm(log(price) ~ log(carat) + cut, data = diamonds)
gg <- ggcoef(m)
# dynamicTicks means generate new axis ticks on zoom
ggplotly(gg, dynamicTicks = TRUE)
```

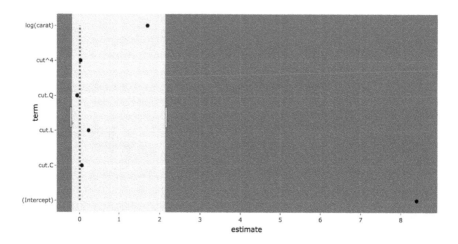

FIGURE 2.13: Zooming in on a coefficient plot generated from the ggcoef() function from the **GGally** package. For a video demonstration of the interactive, see https://bit.ly/GGally. For the interactive, see https://plotly-r.com/interactives/ggally.html

Although the diamonds dataset does not contain any missing values, it's a very common problem in real data analysis problems. The **naniar** package provides a suite of computational and visual resources for working with and revealing structure in missing values. All the **ggplot2** based visualizations return an object that can be converted by ggplotly(). Moreover, **naniar** provides a custom geom, geom_miss_point(), that can be useful for visualizing missingness structure. Figure 2.14 demonstrates this by introducing fake missing values to the diamond price.

```r
library(naniar)
# fake some missing data
diamonds$price_miss <- ifelse(diamonds$depth>60, diamonds$price, NA)
p <- ggplot(diamonds, aes(x = clarity, y = log(price_miss))) +
    geom_miss_point(alpha = 0.1) +
    stat_summary(fun.data = "mean_cl_boot", colour = "black") +
    facet_wrap(~cut)
toWebGL(ggplotly(p))
```

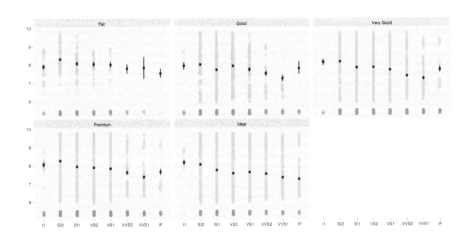

FIGURE 2.14: Using the `geom_miss_point()` function from the **naniar** package to visualize missing values in relation to non-missing values. Missing values are shown in red.

In short, the **ggplot2** ecosystem provides a world-class exploratory visualization toolkit, and having the ability to quickly insert interactivity such as hover, zoom, and filter via `ggplotly()` makes it even more powerful for exploratory analysis. In this introduction to `ggplotly()`, we've only seen relatively simple techniques that come for free out-of-the-box, but the true power of interactive graphics lies in linking multiple views. In that part of the book, you can find lots of examples of linking multiple (`ggplotly()` and `plot_ly()`) graphs purely client-side as well as with **shiny**.

It's also worth mentioning that `ggplotly()` conversions are not always perfect and **ggplot2** doesn't provide an API for interactive features, so sometimes it's desirable to modify the return values of `ggplotly()`. Chapter 33 talks generally about modifying the data structure underlying `ggplotly()` (which, by the way, uses the same a plotly.js figure definition as discussed in Section 2.2). Moreover, Section 25.2 outlines various ways to customize the tooltip that `ggplotly()` produces.

3

Scattered foundations

As we learned in Section 2.2, a plotly.js figure contains one (or more) trace(s), and every trace has a type. The trace type scatter is great for drawing low-level geometries (e.g., points, lines, text, and polygons) and provides the foundation for many add_*() functions (e.g., add_markers(), add_lines(), add_paths(), add_segments(), add_ribbons(), add_area(), and add_polygons()) as well as many ggplotly() charts. These scatter-based layers provide a more convenient interface to special cases of the scatter trace by doing a bit of data wrangling and transformation under-the-hood before mapping to scatter trace(s). For a simple example, add_lines() ensures lines are drawn according to the ordering of x, which is desirable for a time series plotting. This behavior is subtly different than add_paths() which uses row ordering instead.

```r
library(plotly)
data(economics, package = "ggplot2")

# sort economics by psavert, just to
# show difference between paths and lines
p <- economics %>%
  arrange(psavert) %>%
  plot_ly(x = ~date, y = ~psavert)

add_paths(p)
add_lines(p)
```

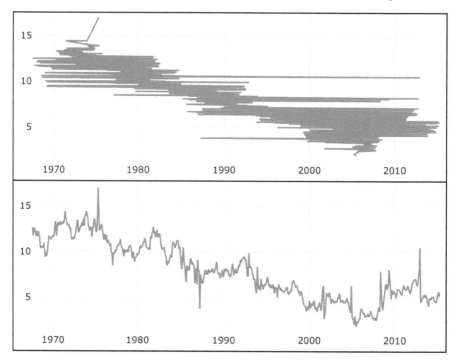

FIGURE 3.1: The difference between `add_paths()` and `add_lines()`: The top panel connects observations according to the ordering of `psavert` (personal savings rate), whereas the bottom panel connects observations according to the ordering of `x` (the date).

Section 2.1 introduced 'aesthetic mapping' arguments (unique to the R package) which make it easier to map data to visual properties (e.g., `color`, `linetype`, etc.). In addition to these arguments, **dplyr** groupings can be used to ensure there is at least one geometry per group. The top panel of Figure 3.1 demonstrates how `group_by()` could be used to effectively wrap the time series from Figure 3.1 by year, which can be useful for visualizing annual seasonality. Another approach to generating at least one geometry per 'group' is to provide categorical variable to a relevant aesthetic (e.g., `color`), as shown in the bottom panel of Figure 3.1.

```
library(lubridate)
econ <- economics %>%
```

```
  mutate(yr = year(date), mnth = month(date))

# One trace (more performant, but less interactive)
econ %>%
  group_by(yr) %>%
  plot_ly(x = ~mnth, y = ~uempmed) %>%
  add_lines(text = ~yr)

# Multiple traces (less performant, but more interactive)
plot_ly(econ, x = ~mnth, y = ~uempmed) %>%
  add_lines(color = ~ordered(yr))

# The split argument guarantees one trace per group level (regardless
# of the variable type). This is useful if you want a consistent
# visual property over multiple traces
# plot_ly(econ, x = ~mnth, y = ~uempmed) %>%
#   add_lines(split = ~yr, color = I("black"))
```

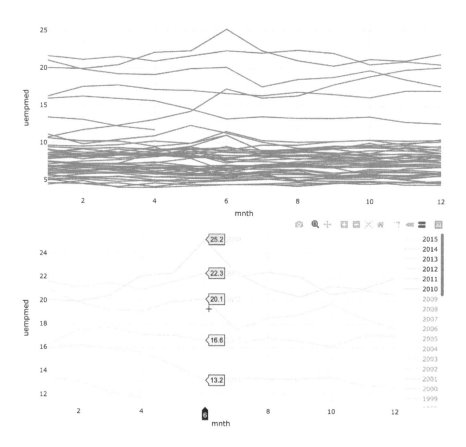

FIGURE 3.2: Drawing multiple lines using **dplyr** groups (top panel) versus a categorical `color` mapping (bottom panel). Comparatively speaking, the bottom panel has more interactive capabilities (e.g., legend-based filtering and multiple tooltips), but it does not scale as well with many lines. For a video demonstration of the interactive, see `https://bit.ly/scatter-lines`. For the interactive, see `https://plotly-r.com/interactives/scatter-lines.html`

Not only do these plots differ in visual appearance, they also differ in interactive capabilities, computational performance, and underlying implementation. That's because, the grouping approach (top panel of Figure 3.2) uses just one plotly.js trace (more performant, less interactive), whereas the `color` approach (bottom panel of Figure 3.2) generates one trace per line/year. In this case, the benefit of having multiple traces is that we can perform interactive filtering via the legend and compare multiple y-values at a given x. The cost of having those capabilities is that plots begin to suffer from performance issues after a few hundred traces, whereas thousands of lines can be rendered fairly easily in one trace. See Chapter 24 for more details on scaling and performance.

These features make it easier to get started using plotly.js, but it still pays off to learn how to use plotly.js directly. You won't find plotly.js attributes listed as explicit arguments in any **plotly** function (except for the special `type` attribute), but they are passed along verbatim to the plotly.js figure definition through the ... operator. The scatter-based layers in this chapter fix the `type` plotly.js attribute to `"scatter"` as well as the `mode`[1] (e.g., `add_markers()` uses `mode='markers'` etc.), but you could also use the lower-level `add_trace()` to work more directly with plotly.js. For example, Figure 3.3 shows how to render markers, lines, and text in the same scatter trace. It also demonstrates how to leverage *nested* plotly.js attributes, like `textfont`[2] and `xaxis`[3]; these attributes contain other attributes, so you need to supply a suitable named list to these arguments.

```r
set.seed(99)
plot_ly() %>%
  add_trace(
    type = "scatter",
    mode = "markers+lines+text",
    x = 4:6,
    y = 4:6,
    text = replicate(3, praise::praise("You are ${adjective}! ")),
    textposition = "right",
    hoverinfo = "text",
    textfont = list(family = "Roboto Condensed", size = 16)
  ) %>%
  layout(xaxis = list(range = c(3, 8)))
```

---

[1]https://plot.ly/r/reference/#scatter-mode
[2]https://plot.ly/r/reference/#scatter-textfont
[3]https://plot.ly/r/reference/#layout-xaxis

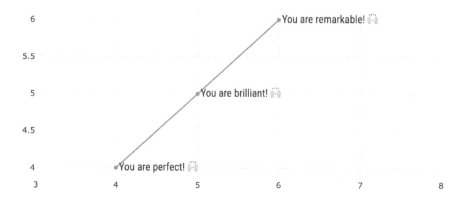

FIGURE 3.3: Using the generic `add_trace()` function to render markers, lines, and text in a single scatter trace. This `add_trace()` function, as well as any `add_*()` function allows you to directly specify plotly.js attributes.

If you are new to plotly.js, I recommend taking a bit of time to look through the plotly.js attributes that are available to the scatter trace type and think how you might be able to use them. Most of these attributes work for other trace types as well, so learning an attribute once for a specific plot can pay off in other contexts as well. The online plotly.js figure reference, `https://plot.ly/r/reference/#scatter`, is a decent place to search and learn about the attributes, but I recommend using the `schema()` function instead for a few reasons:

- `schema()` provides a bit more information than the online docs (e.g., value types, default values, acceptable ranges, etc.).
- The interface makes it a bit easier to traverse and discover new attributes.
- You can be absolutely sure it matches the version used in the R package (the online docs might use a different – probably older – version).

```
schema()
```

```
▼ traces {36}
    ▼ scatter {2}
        ▶ meta {1}
        ▼ attributes {56}
            type : scatter
            ▶ visible {6}
            ▶ showlegend {5}
            ▶ legendgroup {5}
            ▶ opacity {7}
            ▶ name {4}
```

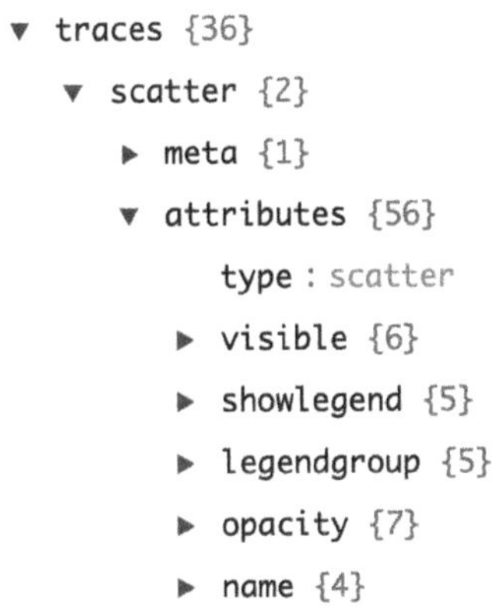

FIGURE 3.4: Using `schema()` function to traverse through the attributes available to a given trace type (e.g., scatter).

The sections that follow in this chapter demonstrate various types of data views using scatter-based layers. In an attempt to avoid duplication of documentation, a particular emphasis is put on features only currently available from the R package (e.g., the aesthetic mapping arguments).

3.1 Markers

This section details scatter traces with a mode of `"markers"` (i.e., `add_markers()`). For simplicity, many of the examples here use `add_markers()` with a numeric x and y axis, which results in scatterplot: a common way to visualize the association between two quantitative variables. The content that follows is still relevant markers displayed non-numeric x and y (aka dot pots) as shown in Section 3.1.6.

3.1.1   Alpha blending

As Unwin (2015) notes, scatterplots can be useful for exposing other important features including: casual relationships, outliers, clusters, gaps, barriers, and conditional relationships. A common problem with scatterplots, however, is overplotting, meaning that there are multiple observations occupying the same (or similar) x/y locations. Figure 3.5 demonstrates one way to combat overplotting via alpha blending. When dealing with tens of thousands of points (or more), consider using toWebGL() to render plots using Canvas rather than SVG (more in Chapter 24), or leveraging 2D density estimation (Section 7.2).

```
subplot(
  plot_ly(mpg, x = ~cty, y = ~hwy, name = "default"),
  plot_ly(mpg, x = ~cty, y = ~hwy) %>%
    add_markers(alpha = 0.2, name = "alpha")
)
```

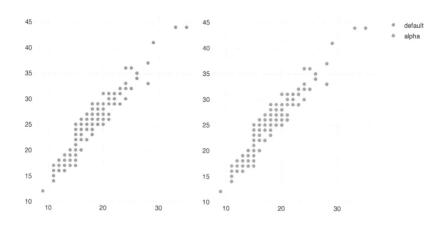

FIGURE 3.5: Combating overplotting in a scatterplot with alpha blending.

3.1.2 Colors

As discussed in Section 2.2, mapping a discrete variable to `color` produces one trace per category, which is desirable for its legend and
hover properties. On the other hand, mapping a *numeric* variable to
`color` produces one trace, as well as a colorbar[4] guide for visually decoding colors back to data values. The `colorbar()` function can be used to
customize the appearance of this automatically generated guide. The
default colorscale is viridis, a perceptually uniform colorscale (even
when converted to black-and-white), and perceivable even to those
with common forms of color blindness (Berkeley Institute for Data
Science, 2016). Viridis is also the default colorscale for ordered factors.

```
p <- plot_ly(mpg, x = ~cty, y = ~hwy, alpha = 0.5)
subplot(
  add_markers(p, color = ~cyl, showlegend = FALSE) %>%
    colorbar(title = "Viridis"),
  add_markers(p, color = ~factor(cyl))
)
```

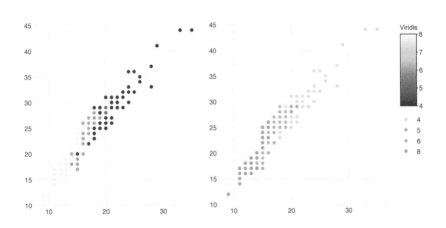

FIGURE 3.6: Variations on a numeric color mapping.

---

[4]https://plot.ly/r/reference/#scatter-marker-colorbar

There are numerous ways to alter the default color scale via the `colors` argument. This argument excepts one of the following: (1) a color brewer palette name (see the row names of `RColor-Brewer::brewer.pal.info` for valid names), (2) a vector of colors to interpolate, or (3) a color interpolation function like `colorRamp()` or `scales::colour_ramp()`. Although this grants a lot of flexibility, one should be conscious of using a sequential colorscale for numeric variables (and ordered factors) as shown in Figure 3.7, and a qualitative colorscale for discrete variables as shown in Figure 3.8.

```
col1 <- c("#132B43", "#56B1F7")
col2 <- viridisLite::inferno(10)
col3 <- colorRamp(c("red", "white", "blue"))
subplot(
  add_markers(p, color = ~cyl, colors = col1) %>%
    colorbar(title = "ggplot2 default"),
  add_markers(p, color = ~cyl, colors = col2) %>%
    colorbar(title = "Inferno"),
  add_markers(p, color = ~cyl, colors = col3) %>%
    colorbar(title = "colorRamp")
) %>% hide_legend()
```

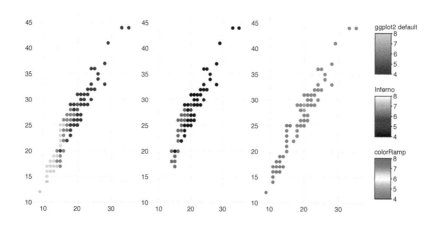

FIGURE 3.7: Three variations on a numeric color mapping.

```
col1 <- "Accent"
col2 <- colorRamp(c("red", "blue"))
col3 <- c(`4` = "red", `5` = "black", `6` = "blue", `8` = "green")
subplot(
  add_markers(p, color = ~factor(cyl), colors = col1),
  add_markers(p, color = ~factor(cyl), colors = col2),
  add_markers(p, color = ~factor(cyl), colors = col3)
) %>% hide_legend()
```

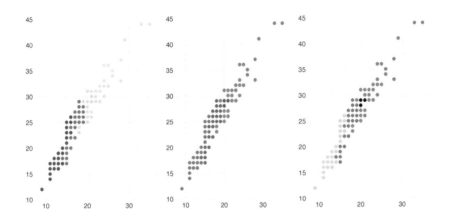

FIGURE 3.8: Three variations on a discrete color mapping.

As introduced in Figure 2.3, color codes can be specified manually (i.e., avoid mapping data values to a visual range) by using the I() function. Figure 3.9 provides a simple example using add_markers(). Any color understood by the col2rgb() function from the **grDevices** package can be used in this way. Chapter 27 provides even more details about working with different color specifications when specifying colors manually.

```
add_markers(p, color = I("black"))
```

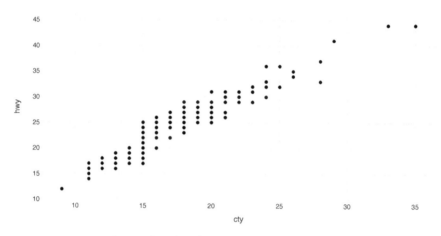

FIGURE 3.9: Setting a fixed color directly using I().

The color argument is meant to control the 'fill-color' of a geometric object, whereas stroke (Section 3.1.4) is meant to control the 'outline-color' of a geometric object. In the case of add_markers(), that means color maps to the plotly.js attribute marker.color[5] and stroke maps to marker.line.color[6]. Not all, but many, marker symbols have a notion of stroke.

3.1.3   Symbols

The symbol argument can be used to map data values to the marker.symbol plotly.js attribute. It uses the same semantics that we've already seen for color:

- A numeric mapping generates trace.
- A discrete mapping generates multiple traces (one trace per category).
- The plural, symbols, can be used to specify the visual range for the mapping.
- Mappings are avoided entirely through I().

For example, the left panel of Figure 3.10 uses a numeric mapping, and the right panel uses a discrete mapping. As a result, the left panel is

---

[5]https://plot.ly/r/reference/#scatter-marker-color
[6]https://plot.ly/r/reference/#scatter-marker-line-color

linked to the first legend entry, whereas the right panel is linked to
the bottom three legend entries. When plotting multiple traces and
no color is specified, the plotly.js colorway[7] is applied (i.e., each trace
will be rendered a different color). To set a fixed color, you can set the
color of every trace generated from this layer with `color = I("black")`,
or similar.

```
p <- plot_ly(mpg, x = ~cty, y = ~hwy, alpha = 0.3)
subplot(
  add_markers(p, symbol = ~cyl, name = "A single trace"),
  add_markers(p, symbol = ~factor(cyl), color = I("black"))
)
```

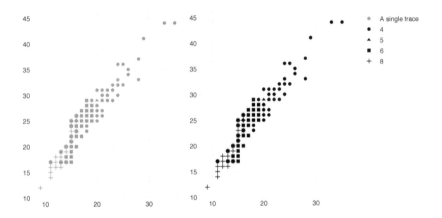

FIGURE 3.10: Mapping symbol to a numeric variable (left panel) and a
factor (right panel).

There are two ways to specify the visual range of `symbols`: (1) numeric
codes (interpreted as a `pch` codes) or (2) a character string specifying a
valid `marker.symbol` value. Figure 3.11 uses pch codes (left panel) as well
as their corresponding `marker.symbol` name (right panel) to specify the
visual range.

---

[7]https://plot.ly/r/reference/#layout-colorway

```
subplot(
  add_markers(p, symbol = ~cyl, symbols = c(17, 18, 19)),
  add_markers(
    p, color = I("black"),
    symbol = ~factor(cyl),
    symbols = c("triangle-up", "diamond", "circle")
  )
)
```

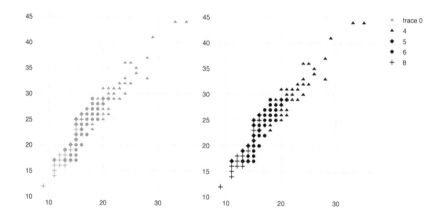

FIGURE 3.11: Specifying the visual range of `symbols`.

These `symbols` (i.e., the visual range) can also be supplied directly to `symbol` through `I()`. For example, Figure 3.12 fixes the marker symbol to a diamond shape.

```
plot_ly(mpg, x = ~cty, y = ~hwy) %>%
  add_markers(symbol = I(18), alpha = 0.5)
```

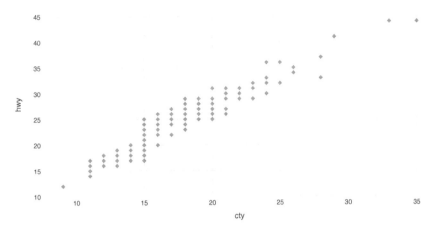

FIGURE 3.12: Setting a fixed symbol directly using I().

If you'd like to see all the symbols available to **plotly**, as well as a method for supplying your own custom glyphs, see Chapter 28.

3.1.4 Stroke and span

The stroke argument follows the same semantics as color and symbol when it comes to variable mappings and specifying visual ranges. Typically you don't want to map data values to stroke, you just want to specify a fixed outline color. For example, Figure 3.13 modifies Figure 3.12 to simply add a black outline. By default, the span, or width of the stroke, is zero, you'll likely want to set the width to be around one pixel.

```
plot_ly(mpg, x = ~cty, y = ~hwy, alpha = 0.5) %>%
  add_markers(symbol = I(18), stroke = I("black"), span = I(1))
```

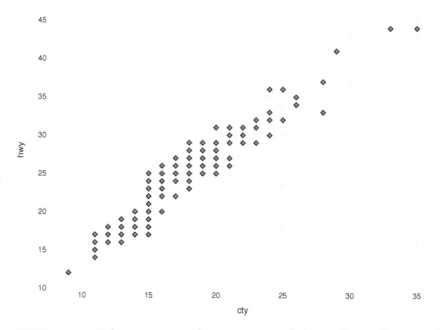

FIGURE 3.13: Using `stroke` and `span` to control the outline color as well as the width of that outline.

3.1.5 Size

For scatterplots, the `size` argument controls the area of markers (unless otherwise specified via sizemode[8]), and *must* be a numeric variable. The `sizes` argument controls the minimum and maximum size of circles, in pixels:

```
p <- plot_ly(mpg, x = ~cty, y = ~hwy, alpha = 0.3)
subplot(
  add_markers(p, size = ~cyl, name = "default"),
  add_markers(p, size = ~cyl, sizes = c(1, 500), name = "custom")
)
```

---

[8] https://plot.ly/r/reference/#scatter-marker-sizemode

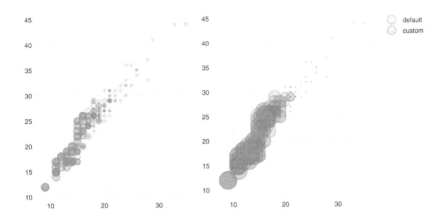

FIGURE 3.14: Controlling the size range via `sizes` (measured in pixels).

Similar to other arguments, `I()` can be used to specify the size directly. In the case of markers, `size` controls the `marker.size`[9] plotly.js attribute. Remember, you always have the option to set this attribute directly by doing something similar to Figure 3.15.

```
plot_ly(mpg, x = ~cty, y = ~hwy, alpha = 0.3, size = I(30))
```

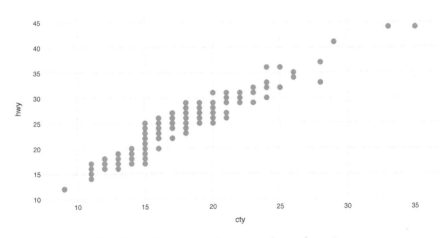

FIGURE 3.15: Setting a fixed marker size directly using `marker.size`.

[9]https://plot.ly/r/reference/#scatter-marker-size

3.1.6   Dotplots and error bars

A dotplot is similar to a scatterplot, except instead of two numeric axes, one is categorical. The usual goal of a dotplot is to compare value(s) on a numerical scale over numerous categories. In this context, dotplots are preferable to pie charts since comparing position along a common scale is much easier than comparing angle or area (Cleveland and McGill, 1984; Heer and Bostock, 2010). Furthermore, dotplots can be preferable to bar charts, especially when comparing values within a narrow range far away from 0 (Few, 2006). Also, when presenting point estimates, and uncertainty associated with those estimates, bar charts tend to exaggerate the difference in point estimates, and lose focus on uncertainty (Messing, 2012).

A popular application for dotplots (with error bars) is the so called "coefficient plot" for visualizing the point estimates of coefficients and their standard error. The `coefplot()` function in the **coefplot** package (Lander, 2016) and the `ggcoef()` function in the **GGally** both produce coefficient plots for many types of model objects in R using **ggplot2**, which we can translate to plotly via `ggplotly()`. Since these packages use points and segments to draw the coefficient plots, the hover information is not the best, and it would be better to use error objects[10]. Figure 3.16 uses the `tidy()` function from the **broom** package (Robinson, 2016) to obtain a data frame with one row per model coefficient, and produce a coefficient plot with error bars along the x-axis.

```
# Fit a full-factorial linear model
m <- lm(
  Sepal.Length ~ Sepal.Width * Petal.Length * Petal.Width,
  data = iris
)


# (1) get a tidy() data structure of covariate-level info
# (e.g., point estimate, standard error, etc.)
# (2) make sure term column is a factor ordered by the estimate
# (3) plot estimate by term with an error bar for the standard error
```

---

[10]`https://plot.ly/r/reference/#scatter-error_x`

```
broom::tidy(m) %>%
  mutate(term = forcats::fct_reorder(term, estimate)) %>%
  plot_ly(x = ~estimate, y = ~term) %>%
  add_markers(
    error_x = ~list(value = std.error),
    color = I("black"),
    hoverinfo = "x"
  )
```

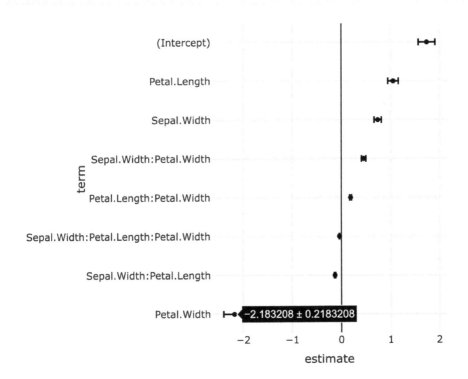

FIGURE 3.16: A coefficient plot.

3.2   Lines

Many of the same principles we learned about aesthetic mappings with
respect to markers (Section 3.1) also apply to lines.[11] Moreover, at the
start of this chapter (namely Figure 3.2) we also learned how to use
dplyr's `group_by()` to ensure there is at least one geometry (in this case,
line) per group. We also learned the difference between `add_paths()`
and `add_lines()`; the former draws lines according to row ordering,
whereas the latter draw them according to x. In this chapter, we'll
learn about `linetype/linetype`, an aesthetic that applies to lines and
polygons. We'll also discuss some other important chart types that can
be implemented with `add_paths()`, `add_lines()`, and `add_segments()`.

3.2.1   Linetypes

Generally speaking, it's hard to perceive more than 8 different col-
ors/linetypes/symbols in a given plot, so sometimes we have to filter
data to use these effectively. Here we use the **dplyr** package to find the
top 5 cities in terms of average monthly sales (`top5`), then effectively
filter the original data to contain just these cities via `semi_join()`. As
Figure 3.17 demonstrates, once we have the data filtered, mapping city
to `color` or `linetype` is trivial. The color palette can be altered via the
`colors` argument, and follows the same rules as scatterplots. The linetype palette can be altered via the `linetypes` argument, and accepts R's
`lty` values[13] or plotly.js dash values[14].

```
library(dplyr)
top5 <- txhousing %>%
```

---

[11]At the time of writing, the plotly.js attributes `line.width` and `line.color`[12] do not
support multiple values, meaning a single line trace can only have one width/color in
2D line plot, and consequently numeric `color/size` mappings won't work. This isn't
necessarily true for 3D paths/lines and there will likely be support for these features
for 2D paths/lines in WebGL in the near future.

[13]https://github.com/wch/r-source/blob/e5b21d/src/library/graphics/man/par.
Rd#L726-L743

[14]https://plot.ly/r/reference/#scatter-line-dash

```
  group_by(city) %>%
  summarise(m = mean(sales, na.rm = TRUE)) %>%
  arrange(desc(m)) %>%
  top_n(5)

tx5 <- semi_join(txhousing, top5, by = "city")

plot_ly(tx5, x = ~date, y = ~median) %>%
  add_lines(linetype = ~city)
```

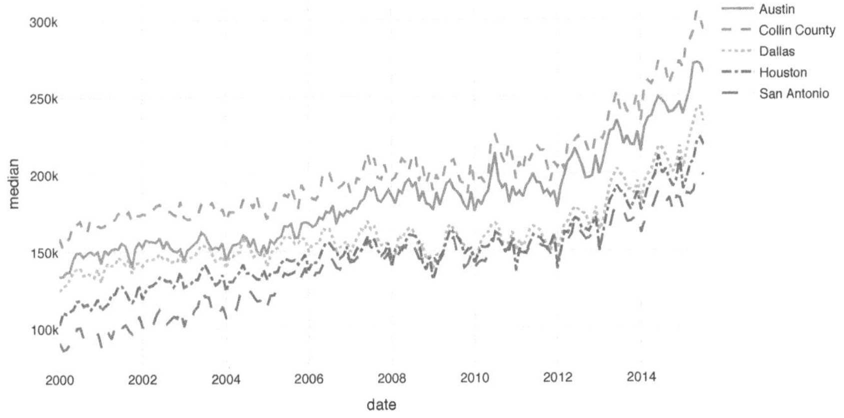

FIGURE 3.17: Using `color` and/or `linetype` to differentiate groups of lines.

If you'd like to control exactly which linetype is used to encode a particular data value, you can provide a named character vector, like in Figure 3.18. Note that this is similar to how we provided a discrete colorscale manually for markers in Figure 3.8.

```
ltys <- c(
  Austin = "dashdot",
  `Collin County` = "longdash",
  Dallas = "dash",
  Houston = "solid",
```

```
  `San Antonio` = "dot"
)

plot_ly(tx5, x = ~date, y = ~median) %>%
  add_lines(linetype = ~city, linetypes = ltys)
```

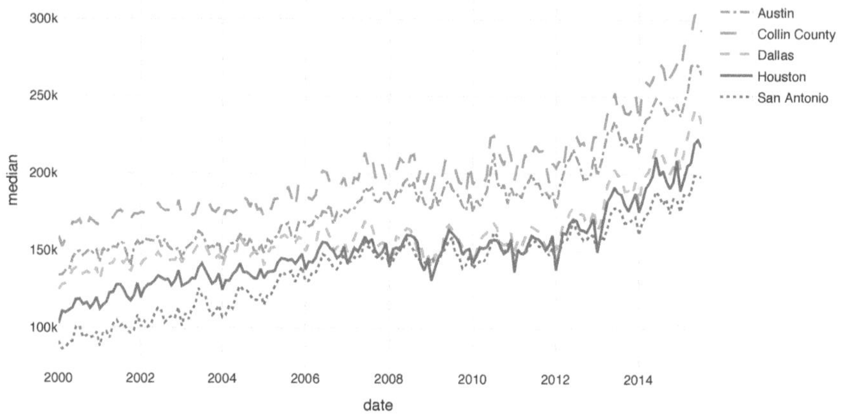

FIGURE 3.18: Providing a named character vector to linetypes in order to control exactly what linetype gets mapped to which city.

3.2.2   Segments

The add_segments() function essentially provides a way to connect two points [(x, y) to (xend, yend)] with a line. Segments form the building blocks for numerous useful chart types, including slopegraphs, dumbell charts, candlestick charts, and more. Slopegraphs and dumbell charts are useful for comparing numeric values across numerous categories. Candlestick charts are typically used for visualizing change in a financial asset over time.

Segments can also provide a useful alternative to add_bars() (covered in Chapter 5), especially for animations. In particular, Figure 14.5 of Section 14.2 shows how to implement an animated population pyramid using segments instead of bars.

3.2.2.1 Slopegraph

!range@range}

The slope graph, made popular by Tufte (2001), is a great way to compare the change in a measurement across numerous groups. This change could be along either a discrete or a continuous axis. For a continuous axis, the slopegraph could be thought of as a decomposition of a line graph into multiple segments. The **slopegraph** R package provides a succinct interface for creating slopegraphs with base or **ggplot2** graphics and also some convenient datasets which we'll make use of here (Leeper, 2017). Figure 3.19 recreates an example from Tufte (2001), using the gdp dataset from **slopegraph**, and demonstrates a common issue with labelling in slopegraphs; it's easy to have overlapping labels when anchoring labels on data values. For that reason, this implementation leverages **plotly** ability to interactively edit annotation positions. See Chapter 12 for similar examples of 'editing views'.

```
data(gdp, package = "slopegraph")
gdp$Country <- row.names(gdp)

plot_ly(gdp) %>%
  add_segments(
    x = 1, xend = 2,
    y = ~Year1970, yend = ~Year1979,
    color = I("gray90")
  ) %>%
  add_annotations(
    x = 1, y = ~Year1970,
    text = ~paste(Country, "  ", Year1970),
    xanchor = "right", showarrow = FALSE
  ) %>%
  add_annotations(
    x = 2, y = ~Year1979,
    text = ~paste(Year1979, "  ", Country),
    xanchor = "left", showarrow = FALSE
  ) %>%
  layout(
```

```
  title = "Current Receipts of Govermnent as a Percentage of GDP",
  showlegend = FALSE,
  xaxis = list(
    range = c(0, 3),
    ticktext = c("1970", "1979"),
    tickvals = c(1, 2),
    zeroline = FALSE
  ),
  yaxis = list(
    title = "",
    showgrid = FALSE,
    showticks = FALSE,
    showticklabels = FALSE
  )
) %>%
config(edits = list(annotationPosition = TRUE))
```

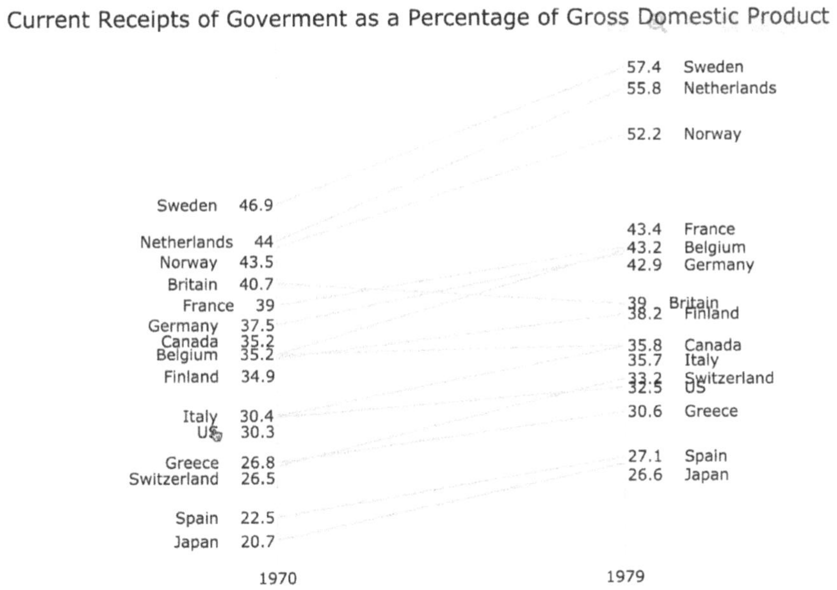

Current Receipts of Goverment as a Percentage of Gross Domestic Product

	57.4	Sweden
	55.8	Netherlands
	52.2	Norway

Sweden   46.9

	43.4	France
Netherlands 44	43.2	Belgium
Norway 43.5	42.9	Germany
Britain 40.7		
France 39	39	Britain
	38.2	Finland
Germany 37.5		
Canada 35.2	35.8	Canada
Belgium 35.2	35.7	Italy
Finland 34.9	33.2	Switzerland
	32.5	US
Italy 30.4	30.6	Greece
US 30.3		
Greece 26.8	27.1	Spain
Switzerland 26.5	26.6	Japan
Spain 22.5		
Japan 20.7		

1970   1979

FIGURE 3.19: Interactively editing the label positioning in a slope-graph. For a video demonstration of the interactive, see https://bit.ly/ Slopegraph. For the interactive, see https://plotly-r.com/interactives/ slopegraph.html

3.2.2.2 Dumbell

So called dumbell charts are similar in concept to slope graphs, but not quite as general. They are typically used to compare two different classes of numeric values across numerous groups. Figure 3.20 uses the dumbell approach to show average miles per gallon city and highway for different car models. With a dumbell chart, it's always a good idea to order the categories by a sensible metric; for Figure 3.20, the categories are ordered by the city miles per gallon.

```
mpg %>%
  group_by(model) %>%
  summarise(c = mean(cty), h = mean(hwy)) %>%
  mutate(model = forcats::fct_reorder(model, c)) %>%
  plot_ly() %>%
  add_segments(
```

```
  x = ~c, y = ~model,
  xend = ~h, yend = ~model,
  color = I("gray"), showlegend = FALSE
) %>%
add_markers(
  x = ~c, y = ~model,
  color = I("blue"),
  name = "mpg city"
) %>%
add_markers(
  x = ~h, y = ~model,
  color = I("red"),
  name  = "mpg highway"
) %>%
layout(xaxis = list(title = "Miles per gallon"))
```

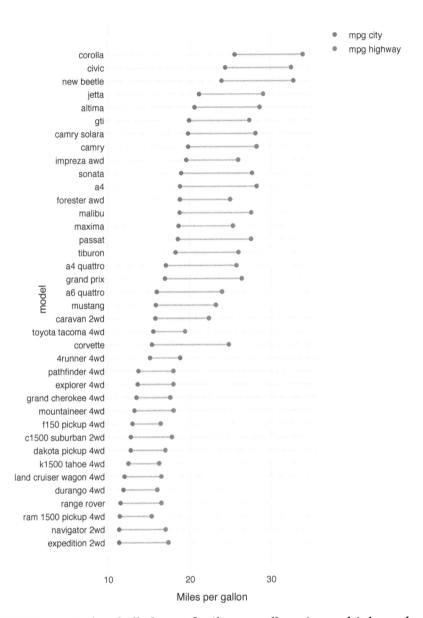

FIGURE 3.20: A dumbell chart of mile per gallon city vs. highway by model of car.

3.2.2.3 Candlestick

Figure 3.21 uses the **quantmod** package (Ryan, 2016) to obtain stock price data for Microsoft and plots two segments for each day: one to encode the opening/closing values, and one to encode the daily high/low. This implementation uses `add_segments()` to implement the candlestick chart, but more recent versions of plotly.js contain a candlestick[15] and ohlc[16] trace types, both of which are useful for visualizing financial data.

```
library(quantmod)
msft <- getSymbols("MSFT", auto.assign = F)
dat <- as.data.frame(msft)
dat$date <- index(msft)
dat <- subset(dat, date >= "2016-01-01")

names(dat) <- sub("^MSFT\\.", "", names(dat))

plot_ly(dat, x = ~date, xend = ~date, color = ~Close > Open,
        colors = c("red", "forestgreen"), hoverinfo = "none") %>%
  add_segments(y = ~Low, yend = ~High, size = I(1)) %>%
  add_segments(y = ~Open, yend = ~Close, size = I(3)) %>%
  layout(showlegend = FALSE, yaxis = list(title = "Price")) %>%
  rangeslider()
```

---

[15] https://plot.ly/r/reference/#candlestick
[16] https://plot.ly/r/reference/#ohlc

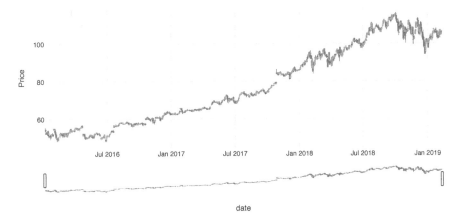

FIGURE 3.21: A candlestick chart built with `add_segments()`. Note how the `color` mapping, which is a logical vector (TRUE if the closing value was higher than opening), creates two traces: a red trace indicating a drop in price and a green trace indicating a rise in price.

3.2.3 Density plots

In Chapter 5, we leverage a number of algorithms in R for computing the "optimal" number of bins for a histogram, via `hist()`, and routing those results to `add_bars()`. We can leverage the `density()` function for computing kernel density estimates in a similar way, and route the results to `add_lines()`, as is done in Figure 3.22.

```
kerns <- c("gaussian", "epanechnikov", "rectangular",
           "triangular", "biweight", "cosine", "optcosine")
p <- plot_ly()
for (k in kerns) {
  d <- density(economics$pce, kernel = k, na.rm = TRUE)
  p <- add_lines(p, x = d$x, y = d$y, name = k)
}
p
```

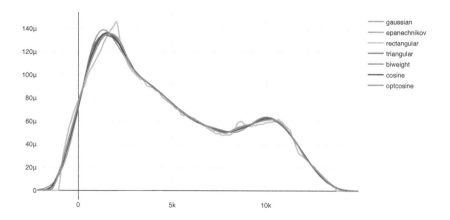

FIGURE 3.22: Various kernel density estimates.

3.2.4 Parallel coordinates

One very useful, but often overlooked, visualization technique is the parallel coordinates plot. Parallel coordinates provide a way to compare values along a common (or non-aligned) positional scale(s) — the most basic of all perceptual tasks — in more than 3 dimensions (Cleveland and McGill, 1984). Usually each line represents every measurement for a given row (or observation) in a dataset. It's true that plotly.js provides a trace type, parcoords, specifically for parallel coordinates that offer desirable interactive capabilities (e.g., highlighting and reordering of axes).[17] However, it can also be useful to learn how to use add_lines() to implement parallel coordinates, as it can offer more flexibility and control over the axis scales.

When measurements are on very different scales, some care must be taken, and variables must be transformed to be put on a common scale. As Figure 3.23 shows, even when variables are measured on a similar scale, it can still be informative to transform variables in different ways.

---

[17]See https://plot.ly/r/parallel-coordinates-plot/ for some interactive examples.

```
iris$obs <- seq_len(nrow(iris))
iris_pcp <- function(transform = identity) {
  iris[] <- purrr::map_if(iris, is.numeric, transform)
  tidyr::gather(iris, variable, value, -Species, -obs) %>%
    group_by(obs) %>%
    plot_ly(x = ~variable, y = ~value, color = ~Species) %>%
    add_lines(alpha = 0.3)
}
subplot(
  iris_pcp(),
  iris_pcp(scale),
  iris_pcp(scales::rescale),
  nrows = 3, shareX = TRUE
) %>% hide_legend()
```

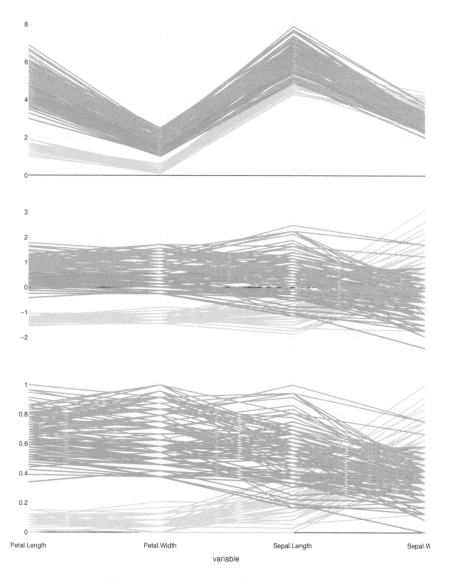

FIGURE 3.23: Parallel coordinate plots of the Iris dataset. The top panel shows all variables on a common scale. The middle panel scales each variable to have mean of 0 and standard deviation of 1. In the bottom panel, each variable is scaled to have a minimum of 0 and a maximum of 1.

It is also worth noting that the **GGally** offers a `ggparcoord()` function which creates parallel coordinate plots via **ggplot2**, which we can convert to plotly via `ggplotly()`. Thanks to the linked highlighting framework, parallel coordinates created in this way could be linked to lower dimensional (but sometimes higher resolution) graphics of related data to guide multi-variate data exploration. The **pedestrians** package provides some examples of linking parallel coordinates to other views such as a grand tour for exposing unusual features in a high-dimensional space (Sievert, 2019a).

3.3 Polygons

The `add_polygons()` function is essentially equivalent to `add_paths()` with the fill[18] attribute set to "toself". Polygons form the basis for other, higher-level scatter-based layers (e.g., `add_ribbons()` and `add_sf()`) that don't have a dedicated plotly.js trace type. Polygons can be used to draw many things, but perhaps the most familiar application where you *might* want to use `add_polygons()` is to draw geo-spatial objects. If and when you use `add_polygons()` to draw a map, make sure you fix the aspect ratio (e.g., `xaxis.scaleanchor`[19]) and also consider using `plotly_empty()` over `plot_ly()` to hide axis labels, ticks, and the background grid. On the other hand, Section 4.2 shows you how to make custom maps using the **sf** package and `add_sf()`, which is a bit of work to get started, but is absolutely worth the investment.

```r
base <- map_data("world", "canada") %>%
  group_by(group) %>%
  plotly_empty(x = ~long, y = ~lat, alpha = 0.2) %>%
  layout(showlegend = FALSE, xaxis = list(scaleanchor = "y"))

base %>%
  add_polygons(hoverinfo = "none", color = I("black")) %>%
```

---

[18] https://plot.ly/r/reference/#scatter-fill
[19] https://plot.ly/r/reference/#layout-xaxis-scaleanchor

```
add_markers(text = ~paste(name, "<br />", pop), hoverinfo = "text",
            color = I("red"), data = maps::canada.cities)
```

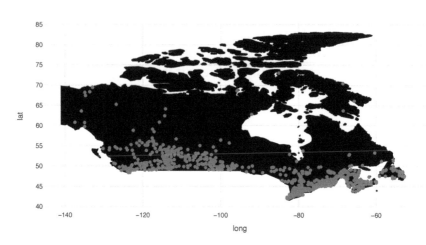

FIGURE 3.24: Using `add_polygons()` to make a map of Canada and major Canadian cities via data provided by the **maps** package.

As discussion surrounding Figure 4.10 points out, scatter-based polygon layers (i.e., `add_polygons()`, `add_ribbons()`, etc.) render all the polygons using one plotly.js trace by default. This approach is computationally efficient, but it's not always desirable (e.g., can't have multiple fills per trace, interactivity is relatively limited). To work around the limitations, consider using `split` (or `color` with a discrete variable) to split the polygon data into multiple traces. Figure 3.25 demonstrates using `split` which will impose plotly.js's colorway to each trace (i.e., subregion) and leverage `hoveron` to generate one tooltip per sub-region.

```
add_polygons(base, split = ~subregion, hoveron = "fills")
```

FIGURE 3.25: Using `split` to render polygons with different fills and interactive properties.

3.3.1 Ribbons

Ribbons are useful for showing uncertainty bounds as a function of x. The `add_ribbons()` function creates ribbons and requires the arguments: `x`, `ymin`, and `ymax`. The `augment()` function from the **broom** package appends observational-level model components (e.g., fitted values stored as a new column `.fitted`) which is useful for extracting those components in a convenient form for visualization. Figure 3.26 shows the fitted values and uncertainty bounds from a linear model object.

```
m <- lm(mpg ~ wt, data = mtcars)
broom::augment(m) %>%
  plot_ly(x = ~wt, showlegend = FALSE) %>%
  add_markers(y = ~mpg, color = I("black")) %>%
  add_ribbons(ymin = ~.fitted - 1.96 * .se.fit,
              ymax = ~.fitted + 1.96 * .se.fit,
              color = I("gray80")) %>%
  add_lines(y = ~.fitted, color = I("steelblue"))
```

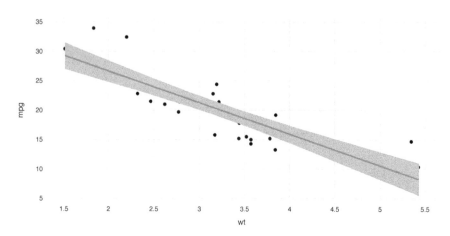

FIGURE 3.26: Plotting fitted values and uncertainty bounds of a linear model via the **broom** package.

4

Maps

There are numerous ways to make a map with **plotly**; each with its own strengths and weaknesses. Generally speaking, the approaches fall under two categories: integrated or custom. Integrated maps leverage plotly.js's built-in support for rendering a basemap layer. Currently there are two supported ways of making integrated maps: either via Mapbox[1] or via an integrated d3.js powered basemap. The integrated approach is convenient if you need a quick map and don't necessarily need sophisticated representations of geo-spatial objects. On the other hand, the custom mapping approach offers complete control since you're providing all the information necessary to render the geo-spatial object(s). Section 4.2 covers making sophisticated maps (e.g., cartograms) using the **sf** R package, but it's also possible to make custom **plotly** maps via other tools for geo-computing (e.g., **sp**, **ggmap**, etc.).

It's worth noting that **plotly** aims to be a general purpose visualization library, and thus, doesn't aim to be the most fully featured geo-spatial visualization toolkit. That said, there are benefits to using **plotly**-based maps since the mapping APIs are very similar to the rest of plotly, and you can leverage the larger **plotly** ecosystem (e.g., linking views client-side like Figure 16.23). However, if you run into limitations with **plotly**'s mapping functionality, there is a very rich set of tools for interactive geospatial visualization in R^2, including but not limited to: **leaflet**, **mapview**, **mapedit**, **tmap**, and **mapdeck** (Lovelace et al., 2019).

---

[1]https://www.mapbox.com/
[2]https://geocompr.robinlovelace.net/adv-map.html#interactive-maps

4.1 Integrated maps

4.1.1 Overview

If you have fairly simple latitude/longitude data and want to make a quick map, you may want to try one of **plotly**'s integrated mapping options (i.e., `plot_mapbox()` and `plot_geo()`). Generally speaking, you can treat these constructor functions as a drop-in replacement for `plot_ly()` and get a dynamic basemap rendered behind your data. Furthermore, all the scatter-based layers we learned about in Section 3 work as you'd expect it to with `plot_ly()`.[3] For example, Figure 4.1 uses `plot_mapbox()` and `add_markers()` to create a bubble chart:

```
plot_mapbox(maps::canada.cities) %>%
  add_markers(
    x = ~long,
    y = ~lat,
    size = ~pop,
    color = ~country.etc,
    colors = "Accent",
    text = ~paste(name, pop),
    hoverinfo = "text"
  )
```

---

[3]Unfortunately, non-scatter traces currently don't work with `plot_mapbox()`/`plot_geo()` meaning that, for one, raster (i.e., heatmap) maps are not natively supported.

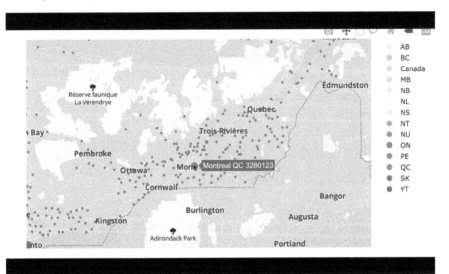

FIGURE 4.1: A mapbox powered bubble chart showing the population of various cities in Canada. For a video demonstration of the interactive, see `https://bit.ly/mapbox-bubble`. For the interactive, see `https://plotly-r.com/interactives/mapbox-bubble.html`

The Mapbox basemap styling is controlled through the `layout.mapbox.style`[4] attribute. The **plotly** package comes with support for 7 different styles, but you can also supply a custom URL to a custom mapbox style[5]. To obtain all the pre-packaged basemap style names, you can grab them from the official plotly.js `schema()`:

```
styles <- schema()$layout$layoutAttributes$mapbox$style$values
styles
#>  [1] "basic"            "streets"
#>  [3] "outdoors"         "light"
#>  [5] "dark"             "satellite"
#>  [7] "satellite-streets" "open-street-map"
#>  [9] "white-bg"         "carto-positron"
```

---

[4]`https://plot.ly/r/reference/#layout-mapbox-style`
[5]`https://docs.mapbox.com/help/tutorials/create-a-custom-style/`

```
#> [11] "carto-darkmatter"   "stamen-terrain"
#> [13] "stamen-toner"        "stamen-watercolor"
```

Any one of these values can be used for a mapbox style. Figure 4.2 demonstrates the satellite earth imagery basemap.

```
layout(
  plot_mapbox(),
  mapbox = list(style = "satellite")
)
```

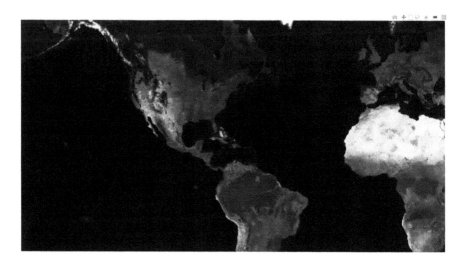

FIGURE 4.2: Zooming in on earth satellite imagery using `plot_mapbox()`. For a video demonstration of the interactive, see `https://bit.ly/mapbox-satellite`. For the interactive, see `https://plotly-r.com/interactives/satellite.html`

Figure 4.3 demonstrates how to create an integrated plotly.js dropdown menu to control the basemap style via the `layout.updatemenus`[6] attribute. The idea behind an integrated plotly.js dropdown is to supply a list of buttons (i.e., menu items) where each button invokes a plotly.js method

---

[6]`https://plot.ly/r/reference/#layout-updatemenus-items-updatemenu-buttons`

with some arguments. In this case, each button uses the relayout[7] method to modify the `layout.mapbox.style` attribute.[8]

```
style_buttons <- lapply(styles, function(s) {
  list(
    label = s,
    method = "relayout",
    args = list("mapbox.style", s)
  )
})
layout(
  plot_mapbox(),
  mapbox = list(style = "dark"),
  updatemenus = list(
    list(y = 0.8, buttons = style_buttons)
  )
)
```

---

[7] https://plot.ly/javascript/plotlyjs-function-reference/

[8] To see more examples of creating and using plotly.js's integrated dropdown functionality to modify graphs, see https://plot.ly/r/dropdowns/

FIGURE 4.3: Providing a dropdown menu to control the styling of the mapbox baselayer. For a video demonstration of the interactive, see `https://bit.ly/mapbox-style-dropdown`. For the interactive, see `https://plotly-r.com/interactives/mapbox-style-dropdown.html`

The other integrated mapping solution in **plotly** is `plot_geo()`. Compared to `plot_mapbox()`, this approach has support for different mapping projections, but styling the basemap is limited and can be more cumbersome. Figure 4.4 demonstrates using `plot_geo()` in conjunction with `add_markers()` and `add_segments()` to visualize flight paths within the United States. Whereas `plot_mapbox()` is fixed to a mercator projection, the `plot_geo()` constructor has a handful of different projections available to it, including the orthographic projection which gives the illusion of the 3D globe.

```
library(plotly)
library(dplyr)
# airport locations
air <- read.csv(
  'https://plotly-r.com/data-raw/airport_locations.csv'
)
# flights between airports
flights <- read.csv(
  'https://plotly-r.com/data-raw/flight_paths.csv'
```

```
)
flights$id <- seq_len(nrow(flights))

# map projection
geo <- list(
  projection = list(
    type = 'orthographic',
    rotation = list(lon = -100, lat = 40, roll = 0)
  ),
  showland = TRUE,
  landcolor = toRGB("gray95"),
  countrycolor = toRGB("gray80")
)


plot_geo(color = I("red")) %>%
  add_markers(
    data = air, x = ~long, y = ~lat, text = ~airport,
    size = ~cnt, hoverinfo = "text", alpha = 0.5
  ) %>%
  add_segments(
    data = group_by(flights, id),
    x = ~start_lon, xend = ~end_lon,
    y = ~start_lat, yend = ~end_lat,
    alpha = 0.3, size = I(1), hoverinfo = "none"
  ) %>%
  layout(geo = geo, showlegend = FALSE)
```

FIGURE 4.4: Using the integrated orthographic projection to visualize flight patterns on a '3D' globe. For a video demonstration of the interactive, see `https://bit.ly/geo-flights`. For the interactive, see `https://plotly-r.com/interactives/geo-flights.html`

One nice thing about `plot_geo()` is that it automatically projects geometries into the proper coordinate system defined by the map projection. For example, in Figure 4.5 the simple line segment is straight when using `plot_mapbox()`, yet curved when using `plot_geo()`. It's possible to achieve the same effect using `plot_ly()` or `plot_mapbox()`, but the relevant marker/line/polygon data has to be put into an **sf** data structure before rendering (see Section 4.2.1 for more details).

```r
map1 <- plot_mapbox() %>%
  add_segments(x = -100, xend = -50, y = 50, yend = 75) %>%
  layout(
    mapbox = list(
      zoom = 0,
      center = list(lat = 65, lon = -75)
    )
  )

map2 <- plot_geo() %>%
  add_segments(x = -100, xend = -50, y = 50, yend = 75) %>%
  layout(geo = list(projection = list(type = "mercator")))

library(htmltools)
browsable(tagList(map1, map2))
```

FIGURE 4.5: A comparison of **plotly**'s integrated mapping solutions: `plot_mapbox()` (top) and `plot_geo()` (bottom). The `plot_geo()` approach will transform line segments to correctly reflect their projection into a non-Cartesian coordinate system.

4.1.2 Choropleths

In addition to scatter traces, both of the integrated mapping solutions (i.e., `plot_mapbox()` and `plot_geo()`) have an optimized choropleth trace type (i.e., the choroplethmapbox[9] and choropleth[10] trace types). Comparatively speaking, choroplethmapbox is more powerful because you can fully specify the feature collection using GeoJSON, but the choropleth trace can be a bit easier to use if it fits your use case.

Figure 4.6 shows the population density of the U.S. via the choropleth trace using the U.S. state data from the **datasets** package (R Core Team, 2016). By simply providing a z[11] attribute, `plotly_geo()` objects will try to create a choropleth, but you'll also need to provide `locations`[12] and a `locationmode`[13]. It's worth noting that the `locationmode` is currently limited to countries and US states, so if you need to plot a different geo-unit (e.g., counties, municipalities, etc.), you should use the choroplethmapbox trace type and/or use a "custom" mapping approach as discussed in Section 4.2.

```r
density <- state.x77[, "Population"] / state.x77[, "Area"]

g <- list(
  scope = 'usa',
  projection = list(type = 'albers usa'),
  lakecolor = toRGB('white')
)

plot_geo() %>%
  add_trace(
    z = ~density, text = state.name, span = I(0),
    locations = state.abb, locationmode = 'USA-states'
  ) %>%
  layout(geo = g)
```

---

[9] https://plot.ly/r/reference/#choroplethmapbox
[10] https://plot.ly/r/reference/#choropleth
[11] https://plot.ly/r/reference/#choropleth-z
[12] https://plot.ly/r/reference/#choropleth-locations
[13] https://plot.ly/r/reference/#choropleth-locationmode

FIGURE 4.6: A map of U.S. population density using the `state.x77` data from the **datasets** package.

Choroplethmapbox is more flexible than choropleth because you supply your own GeoJSON definition of the choropleth via the `geojson` attribute. Currently this attribute must be a URL pointing to a geojson file. Moreover, the `location` should point to a top-level id attribute of each feature within the geojson file. Figure 4.7 demonstrates how we could visualize the same information as Figure 4.6, but this time using choroplethmapbox.

```r
plot_ly() %>%
  add_trace(
    type = "choroplethmapbox",
    # See how this GeoJSON URL was generated at
    # https://plotly-r.com/data-raw/us-states.R
    geojson = paste(c(
      "https://gist.githubusercontent.com/cpsievert/",
      "7cdcb444fb2670bd2767d349379ae886/raw/",
      "cf5631bfd2e385891bb0a9788a179d7f023bf6c8/",
      "us-states.json"
    ), collapse = ""),
    locations = row.names(state.x77),
    z = state.x77[, "Population"] / state.x77[, "Area"],
```

```r
  span = I(0)
) %>%
layout(
  mapbox = list(
    style = "light",
    zoom = 4,
    center = list(lon = -98.58, lat = 39.82)
  )
) %>%
config(
  mapboxAccessToken = Sys.getenv("MAPBOX_TOKEN"),
  # Workaround to make sure image download uses full container
  # size https://github.com/plotly/plotly.js/pull/3746
  toImageButtonOptions = list(
    format = "svg",
    width = NULL,
    height = NULL
  )
)
```

FIGURE 4.7: Another map of U.S. population density, this time using choroplethmapbox with a custom GeoJSON file.

Figures 4.6 and 4.7 aren't an ideal way to visualize state population a graphical perception point of view. We typically use the color in choropleths to encode a numeric variable (e.g., GDP, net exports, average SAT score, etc.) and the eye naturally perceives the area that a particular color covers as proportional to its overall effect. This ends up being misleading since the area the color covers typically has no sensible relationship with the data encoded by the color. A classic example of this misleading effect in action is in US election maps – the proportion of red to blue coloring is not representative of the overall popular vote (Newman, 2016).

Cartograms are an approach to reducing this misleading effect and grant another dimension to encode data through the size of geo-spatial features. Section 4.2.2 covers how to render cartograms in **plotly** using **sf** and **cartogram**.

4.2   Custom maps

4.2.1   Simple features (sf)

The **sf** R package is a modern approach to working with geo-spatial data structures based on tidy data principles (Pebesma, 2018; Wickham, 2014b). The key idea behind **sf** is that it stores geo-spatial geometries in a list-column[14] of a data frame. This allows each row to represent the real unit of observation/interest — whether it's a polygon, multipolygon, point, line, or even a collection of these features — and as a result, works seamlessly inside larger tidy workflows.[15] The **sf** package itself does not really provide geo-spatial data; it provides the framework and utilities for storing and computing on geo-spatial data structures in an opinionated way.

---

[14]`https://jennybc.github.io/purrr-tutorial/ls13_list-columns.html`
[15]This is way more intuitive compared to older workflows based on, say using `ggplot2::fortify()` to obtain a data structure where a row represents a particular point along a feature and having another column track which point belongs to each feature.

There are numerous packages for accessing geo-spatial data as simple features data structures. A couple of notable examples include **rnaturalearth** and **USAboundaries**. The **rnaturalearth** package is better for obtaining any map data in the world via an API provided by https://www.naturalearthdata.com/ (South, 2017). The **USAboundaries** package is great for obtaining map data for the United States at any point in history (Mullen and Bratt, 2018). It doesn't really matter what tool you use to obtain or create an **sf** object; once you have one, plot_ly() knows how to render it:

```
library(rnaturalearth)
world <- ne_countries(returnclass = "sf")
class(world)
#> [1] "sf"     "data.frame"
plot_ly(world, color = I("gray90"), stroke = I("black"), span = I(1))
```

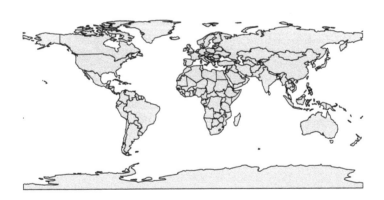

FIGURE 4.8: Rendering all the world's countries using plot_ly() and the ne_countries() function from the **rnaturalearth** package.

How does plot_ly() know how to render the countries? It's because the geo-spatial features are encoded in special (geometry) list-column. Also, meta-data about the geo-spatial structure are retained as special attributes of the data. Figure 4.9 augments the print method for **sf** to data frames to demonstrate that all the information needed to render the countries (i.e., polygons) in Figure 4.8 is contained within the world data frame. Note also that **sf** provides special **dplyr** methods for this special class of data frame so that you can treat data manipulation as if it were a 'tidy' data structure. One thing about this method is that the special 'geometry' column is always retained; if we try to just select the name column, then we get both the name and the geometry.

```
library(sf)
world %>%
  select(name) %>%
  print(n = 4)
```

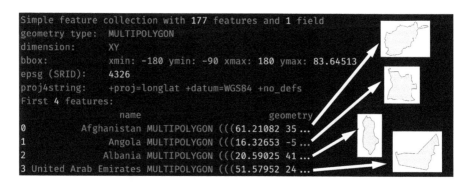

FIGURE 4.9: A diagram of a simple features data frame. The geometry column tracks the spatial features attached to each row in the data frame.

There are actually 4 different ways to render **sf** objects with **plotly**: plot_ly(), plot_mapbox(), plot_geo(), and via **ggplot2**'s geom_sf(). These functions render multiple polygons using a *single* trace by default, which is fast, but you may want to leverage the added flexibility of multiple traces. For example, a given trace can only have one fillcolor, so it's impossible to render multiple polygons with different colors using a single trace. For this reason, if you want to vary the color of multiple polygons, make sure the split by a unique identifier (e.g., name), as done in Figure 4.10. Note that, as discussed for line charts in Figure 3.2, using multiple traces automatically adds the ability to filter name via legend entries.

```
canada <- ne_states(country = "Canada", returnclass = "sf")
plot_ly(canada, split = ~name, color = ~provnum_ne)
```

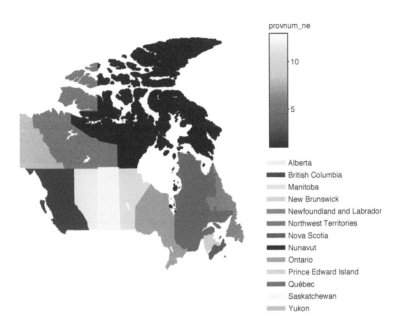

FIGURE 4.10: Using `split` and `color` to create a choropleth map of provinces in Canada.

Another important feature for maps that may require you to `split` multiple polygons into multiple traces is the ability to display a different hover-on-fill for each polygon. By providing `text` that is unique within each polygon and specifying `hoveron='fills'`, as in Figure 4.11, the tooltip behavior is tied to the trace's fill (instead of being displayed at each point along the polygon).

```
plot_ly(
  canada,
  split = ~name,
  color = I("gray90"),
  text = ~paste(name, "is \n province number", provnum_ne),
```

```
  hoveron = "fills",
  hoverinfo = "text",
  showlegend = FALSE
)
```

FIGURE 4.11: Using `split`, `text`, and `hoveron='fills'` to display a tooltip specific to each Canadian province.

Although the integrated mapping approaches (`plot_mapbox()` and `plot_geo()`) can render **sf** objects, the custom mapping approaches (`plot_ly()` and `geom_sf()`) are more flexible because they allow for any well-defined mapping projection. Working with and understanding map projections can be intimidating for a causal map maker. Thankfully, there are nice resources for searching map projections in a human-friendly interface, like `http://spatialreference.org/`. Through this website, one can search desirable projections for a given portion of the globe and extract commands for projecting their geo-spatial objects into that projection. As shown in Figure 4.12, one way to perform the projection is to supply the relevant PROJ4 command to the st_transform() function in **sf** (PROJ contributors, 2018).

```
# filter the world sf object down to canada
canada <- filter(world, name == "Canada")
# coerce cities lat/long data to an official sf object
cities <- st_as_sf(
  maps::canada.cities,
  coords = c("long", "lat"),
  crs = 4326
)

# A PROJ4 projection designed for Canada
# http://spatialreference.org/ref/sr-org/7/
# http://spatialreference.org/ref/sr-org/7/proj4/
moll_proj <- "+proj=moll +lon_0=0 +x_0=0 +y_0=0 +ellps=WGS84
+units=m +no_defs"

# perform the projections
canada <- st_transform(canada, moll_proj)
cities <- st_transform(cities, moll_proj)

# plot with geom_sf()
p <- ggplot() +
  geom_sf(data = canada) +
  geom_sf(data = cities, aes(size = pop), color = "red", alpha = 0.3)
ggplotly(p)
```

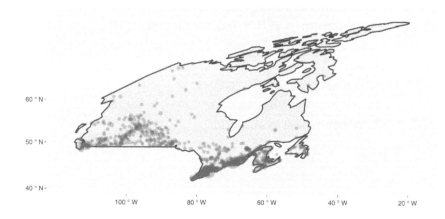

60 ° N -

50 ° N -

40 ° N -

                100 ° W   80 ° W   60 ° W   40 ° W   20 ° W

FIGURE 4.12: The population of various Canadian cities rendered on a custom basemap using a Mollweide projection.

Some geo-spatial objects have an unnecessarily high resolution for a given visualization. In these cases, you may want to consider simplifying the geo-spatial object to improve the speed of the R code and responsiveness of the visualization. For example, we could recreate Figure 4.8 with a much higher resolution by specifying scale = "large" in ne_countries(); this gives us a **sf** object with over 50 times more spatial coordinates than the default scale. The higher resolution allows us to zoom in better on more complex geo-spatial regions, but it allow leads to slower R code, larger HTML files, and slower responsiveness. Sievert (2018b) explores this issue in more depth and demonstrates how to use the st_simplify() function from **sf** to simplify features before plotting them.

```
sum(rapply(world$geometry, nrow))
#> [1] 10586
```

```
world_large <- ne_countries(scale = "large", returnclass = "sf")
sum(rapply(world_large$geometry, nrow))
#> [1] 548121
```

Analogous to the discussion surrounding 3.2, it pays to be aware of the tradeoffs involved with rendering **plotly** graphics using one or many traces, and to be knowledgeable about how to leverage either approach. Specifically, by default, **plotly** attempts to render all simple features in a single trace, which is performant, but doesn't have a lot of interactivity.

```
plot_mapbox(
  world_large,
  color = NA,
  stroke = I("black"),
  span = I(0.5)
)
```

For those interested in learning more about geocomputation in R with **sf** and other great R packages like **sp** and **raster**, Lovelace et al. (2019) provide lots of nice and freely available learning resources (Pebesma and Bivand, 2005; Hijmans, 2019).

4.2.2 Cartograms

Cartograms distort the size of geo-spatial polygons to encode a numeric variable other than the land size. There are numerous types of cartograms and they are typically categorized by their ability to preserve shape and maintain contiguous regions. Cartograms have been shown to be an effective approach to both encode and teach about geo-spatial data, though the effects certainly vary by cartogram type (Nusrat et al., 2016). The R package **cartogram** provides an interface to several popular cartogram algorithms (Jeworutzki, 2018). A number of other R packages provide cartogram algorithms, but the great thing about **cartogram** is that all the functions can take an **sf** (or **sp**) object as input and return an **sf** object. This makes it incredibly easy to go from raw spatial objects, to transformed objects, to visual. Figure 4.13 demonstrates a continuous area cartogram of US population in 2014 using a rubber sheet distortion algorithm from Dougenik et al. (1985).

```
library(cartogram)
library(albersusa)

us_cont <- cartogram_cont(usa_sf("laea"), "pop_2014")

plot_ly(us_cont) %>%
  add_sf(
    color = ~pop_2014,
    split = ~name,
    span = I(1),
    text = ~paste(name, scales::number_si(pop_2014)),
    hoverinfo = "text",
    hoveron = "fills"
  ) %>%
  layout(showlegend = FALSE) %>%
  colorbar(title = "Population \n 2014")
```

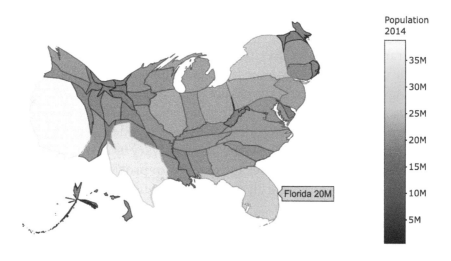

FIGURE 4.13: A cartogram of the U.S. population in 2014. A cartogram sizes the area of geo-spatial objects proportional to some metric (e.g., population).

Figure 4.14 demonstrates a non-continuous Dorling cartogram of US population in 2014 from Dorling D. (1996). This cartogram does not try to preserve the shape of polygons (i.e., states), but instead uses circles to represent each geo-spatial object, then encodes the variable of interest (i.e., population) using the area of the circle.

```r
us <- usa_sf("laea")
us_dor <- cartogram_dorling(us, "pop_2014")

plot_ly(stroke = I("black"), span = I(1)) %>%
  add_sf(
    data = us,
    color = I("gray95"),
    hoverinfo = "none"
  ) %>%
  add_sf(
    data = us_dor,
    color = ~pop_2014,
    split = ~name,
    text = ~paste(name, scales::number_si(pop_2014)),
    hoverinfo = "text",
    hoveron = "fills"
  ) %>%
  layout(showlegend = FALSE)
```

FIGURE 4.14: A Dorling cartogram of the U.S. population in 2014. A Dorling cartogram sizes the circles proportional to some metric (e.g., population).

Figure 4.15 demonstrates a non-continuous cartogram of the U.S. population in 2014 from Olson (1976). In contrast to the Dorling cartogram, this approach does preserve the shape of polygons. The implementation behind Figure 4.15 is to simply take the implementation of Figure 4.14 and change `cartogram_dorling()` to `cartogram_ncont()`.

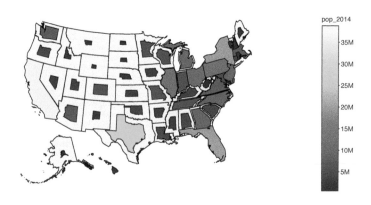

FIGURE 4.15: A non-contiguous cartogram of the U.S. population in 2014 that preserves shape.

A popular class of contiguous cartograms that do not preserve shape are sometimes referred to as tile cartograms (aka tilegrams). At the time of writing, there doesn't seem to be a great R package for *computing* tilegrams, but Pitch Interactive provides a nice web service where you can generate tilegrams from existing or custom data https://pitchinteractiveinc.github.io/tilegrams/. Moreover, the service allows you to download a TopoJSON file of the generated tilegram, which we can read in R and convert into an **sf** object via **geojsonio** (Chamberlain and Teucher, 2018). Figure 4.16 demonstrates a tilegram of U.S. Population in 2016 exported directly from Pitch's free web service.

```
library(geojsonio)
tiles <- geojson_read("~/Downloads/tiles.topo.json", what = "sp")
tiles_sf <- st_as_sf(tiles)
plot_ly(
  tiles_sf, color = ~name,
  colors = "Paired", stroke = I("transparent")
)
```

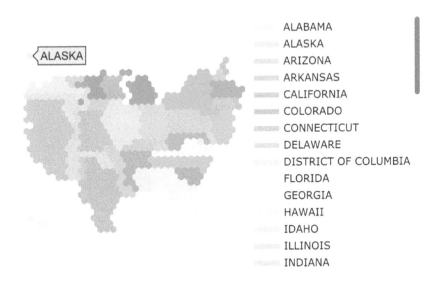

FIGURE 4.16: A tile cartogram of the U.S. population in 2016.

5

Bars and histograms

The `add_bars()` and `add_histogram()` functions wrap the bar[1] and histogram[2] plotly.js trace types. The main difference between them is that bar traces require bar heights (both x and y), whereas histogram traces require just a single variable, and plotly.js handles binning in the browser.[3] And perhaps confusingly, both of these functions can be used to visualize the distribution of either a numeric or a discrete variable. So, essentially, the only difference between them is where the binning occurs.

Figure 5.1 compares the default binning algorithm in plotly.js to a few different algorithms available in R via the `hist()` function. Although plotly.js has the ability to customize histogram bins via `xbins`[4]/`ybins`[5], R has diverse facilities for estimating the optimal number of bins in a histogram that we can easily leverage.[6] The `hist()` function alone allows us to reference 3 famous algorithms by name (Sturges, 1926; Freedman and Diaconis, 1981; Scott, 1979), but there are also packages (e.g., the **histogram** package) which extend this interface to incorporate more methodology (Mildenberger et al., 2009). The `price_hist()` function below wraps the `hist()` function to obtain the binning results, and map those bins to a plotly version of the histogram using `add_bars()`.

---

[1] https://plot.ly/r/reference/#bar

[2] https://plot.ly/r/reference/#histogram

[3] As we'll see in Section 16.1, and specifically Figure 16.6, using a 'statistical' trace type like `add_histogram()` enables statistical graphical queries.

[4] https://plot.ly/r/reference/#histogram-xbins

[5] https://plot.ly/r/reference/#histogram-ybins

[6] Optimal in this context is the number of bins which minimizes the distance between the empirical histogram and the underlying density.

```
p1 <- plot_ly(diamonds, x = ~price) %>%
  add_histogram(name = "plotly.js")

price_hist <- function(method = "FD") {
  h <- hist(diamonds$price, breaks = method, plot = FALSE)
  plot_ly(x = h$mids, y = h$counts) %>% add_bars(name = method)
}

subplot(
  p1, price_hist(), price_hist("Sturges"), price_hist("Scott"),
  nrows = 4, shareX = TRUE
)
```

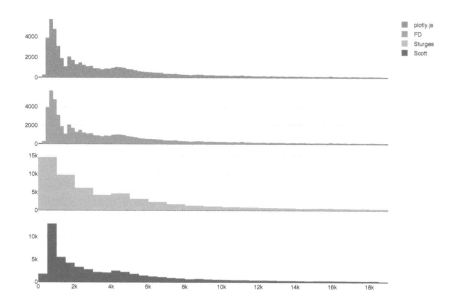

FIGURE 5.1: plotly.js's default binning algorithm versus R's `hist()` default.

Figure 5.2 demonstrates two ways of creating a basic bar chart. Although the visual results are the same, it is worth noting the difference in implementation. The `add_histogram()` function sends all of the observed values to the browser and lets plotly.js perform the binning. It takes more human effort to perform the binning in R, but doing so has the benefit of sending less data, and requiring less computation work of the web browser. In this case, we have only about 50,000 records, so there is not much of a difference in page load times or page size. However, with 1 million records, page load time more than doubles and page size nearly doubles.[7]

```
library(dplyr)
p1 <- plot_ly(diamonds, x = ~cut) %>%
  add_histogram()

p2 <- diamonds %>%
  count(cut) %>%
  plot_ly(x = ~cut, y = ~n) %>%
  add_bars()

subplot(p1, p2) %>% hide_legend()
```

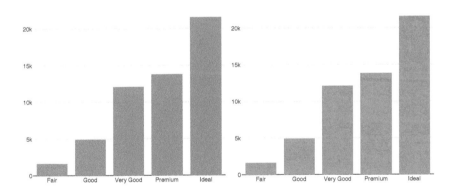

FIGURE 5.2: Number of diamonds by cut.

---

[7]These tests were run on Google Chrome and loaded a page with a single bar chart. See https://www.webpagetest.org/result/160924_DP_JBX for add_histogram() and https://www.webpagetest.org/result/160924_QG_JA1 for add_bars().

5.1   Multiple numeric distributions

It is often useful to see how the numeric distribution changes with
respect to a discrete variable. When using bars to visualize multiple
numeric distributions, I recommend plotting each distribution on its
own axis using a small multiples display, rather than trying to overlay
them on a single axis.[8] Chapter 13, and specifically Section 13.1.2.3,
discusses small multiples in more detail, but Figure 13.9 demonstrates
how it is done with `plot_ly()` and `subplot()`. Note how the `one_plot()`
function defines what to display on each panel, then a split-apply-
recombine (i.e., `split()`, `lapply()`, `subplot()`) strategy is employed to
generate the trellis display.

```
one_plot <- function(d) {
  plot_ly(d, x = ~price) %>%
    add_annotations(
      ~unique(clarity), x = 0.5, y = 1,
      xref = "paper", yref = "paper", showarrow = FALSE
    )
}

diamonds %>%
  split(.$clarity) %>%
  lapply(one_plot) %>%
  subplot(nrows = 2, shareX = TRUE, titleX = FALSE) %>%
  hide_legend()
```

---

[8] It's much easier to visualize multiple numeric distributions on a single axis using
lines.

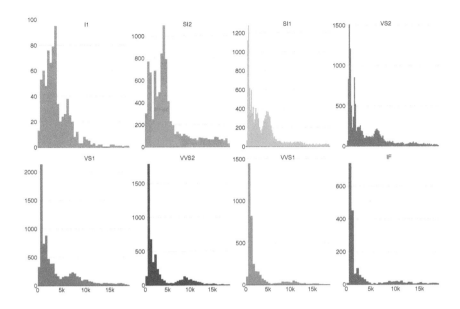

FIGURE 5.3: A trellis display of diamond price by diamond clarity.

5.2   Multiple discrete distributions

Visualizing multiple discrete distributions is difficult. The subtle complexity is due to the fact that both counts and proportions are important for understanding multi-variate discrete distributions. Figure 5.4 presents diamond counts, divided by both their cut and clarity, using a grouped bar chart.

```
plot_ly(diamonds, x = ~cut, color = ~clarity) %>%
  add_histogram()
```

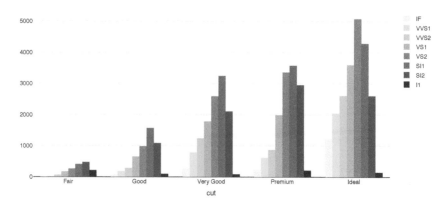

FIGURE 5.4: A grouped bar chart of diamond counts by cut and clarity.

Figure 5.4 is useful for comparing the number of diamonds by clarity, given a type of cut. For instance, within "Ideal" diamonds, a cut of "VS1" is most popular, "VS2" is second most popular, and "I1" the least popular. The distribution of clarity within "Ideal" diamonds seems to be fairly similar to other diamonds, but it's hard to make this comparison using raw counts. Figure 5.5 makes this comparison easier by showing the relative frequency of diamonds by clarity, given a cut.

```
# number of diamonds by cut and clarity (n)
cc <- count(diamonds, cut, clarity)
# number of diamonds by cut (nn)
cc2 <- left_join(cc, count(cc, cut, wt = n, name = 'nn'))
cc2 %>%
  mutate(prop = n / nn) %>%
  plot_ly(x = ~cut, y = ~prop, color = ~clarity) %>%
  add_bars() %>%
  layout(barmode = "stack")
```

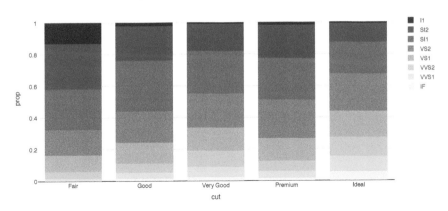

FIGURE 5.5: A stacked bar chart showing the proportion of diamond clarity within cut.

This type of plot, also known as a spine plot, is a special case of a mosaic plot. In a mosaic plot, you can scale both bar widths and heights according to discrete distributions. For mosaic plots, I recommend using the **ggmosaic** package (Jeppson et al., 2016), which implements a custom **ggplot2** geom designed for mosaic plots, which we can convert to plotly via `ggplotly()`. Figure 5.6 shows a mosaic plot of cut by clarity. Notice how the bar widths are scaled proportional to the cut frequency.

```
library(ggmosaic)
p <- ggplot(data = cc) +
  geom_mosaic(aes(weight = n, x = product(cut), fill = clarity))
ggplotly(p)
```

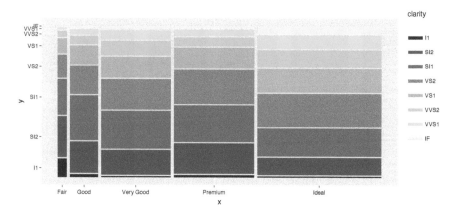

FIGURE 5.6: Using **ggmosaic** and `ggplotly()` to create advanced inter-active visualizations of categorical data.

6

Boxplots

Boxplots encode the five number summary of a numeric variable, and provide a decent way to compare many numeric distributions. The visual task of comparing multiple boxplots is relatively easy (i.e., compare position along a common scale) compared to some common alternatives (e.g., a trellis display of histograms, like Figure 5.1), but the boxplot is sometimes inadequate for capturing complex (e.g., multi-modal) distributions (in this case, a frequency polygon, like Figure 2.9 provides a nice alternative). The add_boxplot() function requires one numeric variable, and guarantees boxplots are oriented[1] correctly, regardless of whether the numeric variable is placed on the x or y scale. As Figure 6.1 shows, on the axis orthogonal to the numeric axis, you can provide a discrete variable (for conditioning) or supply a single value (to name the axis category).

```
p <- plot_ly(diamonds, y = ~price, color = I("black"),
             alpha = 0.1, boxpoints = "suspectedoutliers")
p1 <- p %>% add_boxplot(x = "Overall")
p2 <- p %>% add_boxplot(x = ~cut)
subplot(
  p1, p2, shareY = TRUE,
  widths = c(0.2, 0.8), margin = 0
) %>% hide_legend()
```

---

[1]https://plot.ly/r/reference/#box-orientation

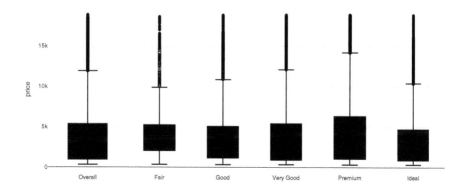

FIGURE 6.1: Overall diamond price and price by cut.

If you want to partition by more than one discrete variable, you could use the interaction of those variables to the discrete axis, and coloring by the nested variable, as Figure 6.2 does with diamond clarity and cut. Another approach would be to use a trellis display, similar to Figure 13.9.

```
plot_ly(diamonds, x = ~price, y = ~interaction(clarity, cut)) %>%
  add_boxplot(color = ~clarity) %>%
  layout(yaxis = list(title = ""))
```

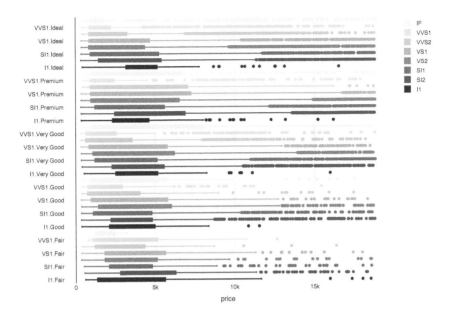

FIGURE 6.2: Diamond prices by cut and clarity.

It is also helpful to sort the boxplots according to something meaningful, such as the median price. Figure 6.3 presents the same information as Figure 6.2, but sorts the boxplots by their median, and makes it immediately clear that diamonds with a cut of "SI2" have the highest diamond price, on average.

```
d <- diamonds %>%
  mutate(cc = interaction(clarity, cut))

# interaction levels sorted by median price
lvls <- d %>%
  group_by(cc) %>%
  summarise(m = median(price)) %>%
  arrange(m) %>%
  pull(cc)

plot_ly(d, x = ~price, y = ~factor(cc, lvls)) %>%
```

```
add_boxplot(color = ~clarity) %>%
layout(yaxis = list(title = ""))
```

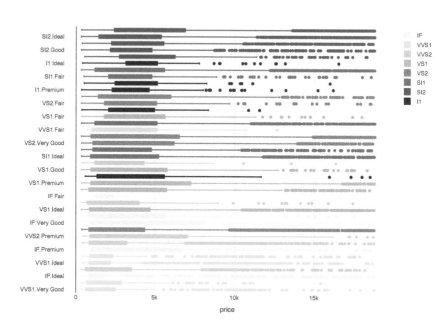

FIGURE 6.3: Diamond prices by cut and clarity, sorted by price median.

Similar to `add_histogram()`, `add_boxplot()` sends the raw data to the browser, and lets plotly.js compute summary statistics. Unfortunately, plotly.js does not yet allow precomputed statistics for boxplots.[2]

---

[2] Follow the issue here https://github.com/plotly/plotly.js/issues/1059

7

2D frequencies

7.1 Rectangular binning in plotly.js

The **plotly** package provides two functions for displaying rectangular bins: `add_heatmap()` and `add_histogram2d()`. For numeric data, the `add_heatmap()` function is a 2D analog of `add_bars()` (bins must be pre-computed), and the `add_histogram2d()` function is a 2D analog of `add_histogram()` (bins can be computed in the browser). Thus, I recommend `add_histogram2d()` for exploratory purposes, since you don't have to think about how to perform binning. It also provides a useful `zsmooth`[1] attribute for effectively increasing the number of bins (currently, "best" performs a bi-linear interpolation[2], a type of nearest neighbors algorithm), and `nbinsx`[3]/`nbinsy`[4] attributes to set the number of bins in the x and/or y directions. Figure 7.1 compares three different uses of `add_histogram()`: (1) plotly.js's default binning algorithm, (2) the default plus smoothing, (3) setting the number of bins in the x and y directions. It is also worth noting that filled contours, instead of bins, can be used in any of these cases by using `add_histogram2dcontour()` instead of `add_histogram2d()`.

```
p <- plot_ly(diamonds, x = ~log(carat), y = ~log(price))
subplot(
  add_histogram2d(p) %>%
    colorbar(title = "default") %>%
    layout(xaxis = list(title = "default")),
```

---

[1]https://plot.ly/r/reference/#histogram2d-zsmooth
[2]https://en.wikipedia.org/wiki/Bilinear_interpolation
[3]https://plot.ly/r/reference/#histogram2d-nbinsx
[4]https://plot.ly/r/reference/#histogram2d-nbinsy

```
add_histogram2d(p, zsmooth = "best") %>%
  colorbar(title = "zsmooth") %>%
  layout(xaxis = list(title = "zsmooth")),
add_histogram2d(p, nbinsx = 60, nbinsy = 60) %>%
  colorbar(title = "nbins") %>%
  layout(xaxis = list(title = "nbins")),
shareY = TRUE, titleX = TRUE
)
```

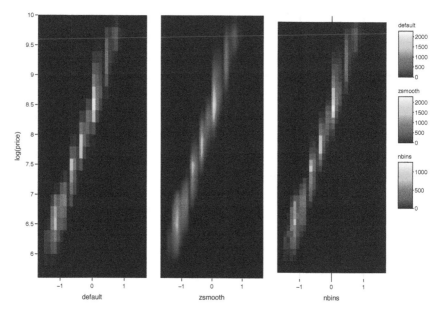

FIGURE 7.1: Three different uses of `histogram2d()`.

7.2   Rectangular binning in R

In Chapter 5, we leveraged a number of algorithms in R for computing the "optimal" number of bins for a histogram, via `hist()`, and routing those results to `add_bars()`. There is a surprising lack of research and computational tools for the 2D analog, and among the research that does exist, solutions usually depend on characteristics of the unknown underlying distribution, so the typical approach is to assume a Gaussian form (Scott, 1992). Practically speaking, that assumption is not very useful, but 2D kernel density estimation provides a useful alternative that tends to be more robust to changes in distributional form. Although kernel density estimation requires choice of kernel and a bandwidth parameter, the kde2d() function from the **MASS** package provides a well-supported rule-of-thumb for estimating the bandwidth of a Gaussian kernel density (Venables and Ripley, 2002). Figure 7.2 uses kde2d() to estimate a 2D density, scales the relative frequency to an absolute frequency, then uses the add_heatmap() function to display the results as a heatmap.

```
kde_count <- function(x, y, ...) {
  kde <- MASS::kde2d(x, y, ...)
  df <- with(kde, setNames(expand.grid(x, y), c("x", "y")))
  # The 'z' returned by kde2d() is a proportion,
  # but we can scale it to a count
  df$count <- with(kde, c(z) * length(x) * diff(x)[1] * diff(y)[1])
  data.frame(df)
}


kd <- with(diamonds, kde_count(log(carat), log(price), n = 30))
plot_ly(kd, x = ~x, y = ~y, z = ~count) %>%
  add_heatmap() %>%
  colorbar(title = "Number of diamonds")
```

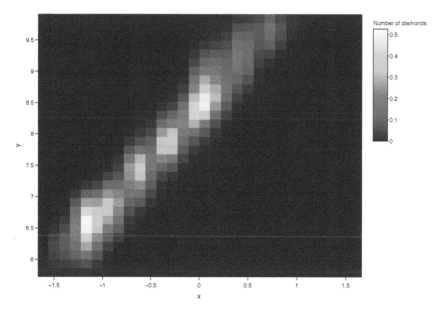

FIGURE 7.2: 2D density estimation via the `kde2d()` function.

7.3   Categorical axes

The functions `add_histogram2d()`, `add_histogram2dcontour()`, and `add_heatmap()` all support categorical axes. Thus, `add_histogram2d()` *can* be used to easily display 2-way contingency tables, but since it is easier to compare values along a common scale rather than compare colors (Cleveland and McGill, 1984), I recommend creating grouped bar charts instead. The `add_heatmap()` function can still be useful for categorical axes, however, as it allows us to display whatever quantity we want along the z axis (color).

Figure 7.3 uses `add_heatmap()` to display a correlation matrix. Notice how the `limits` arguments in the `colorbar()` function can be used to expand the limits of the color scale to reflect the range of possible correlations (something that is not easily done in plotly.js).

```
corr <- cor(dplyr::select_if(diamonds, is.numeric))
plot_ly(colors = "RdBu") %>%
  add_heatmap(x = rownames(corr), y = colnames(corr), z = corr) %>%
  colorbar(limits = c(-1, 1))
```

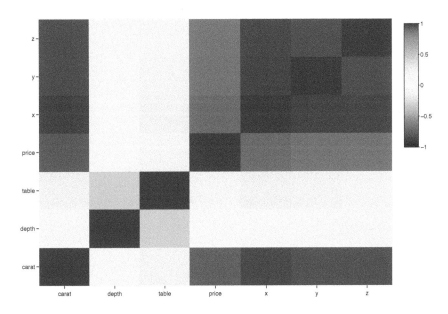

FIGURE 7.3: Displaying a correlation matrix with `add_heatmap()` and controlling the scale limits with `colorbar()`.

8

3D charts

8.1 Markers

As it turns out, by simply adding a z attribute plot_ly() automatically renders markers, lines, and paths in three dimensions. That means, all the techniques we learned in Sections 3.1 and 3.2 can be re-used for 3D charts:

```
plot_ly(mpg, x = ~cty, y = ~hwy, z = ~cyl) %>%
    add_markers(color = ~cyl)
```

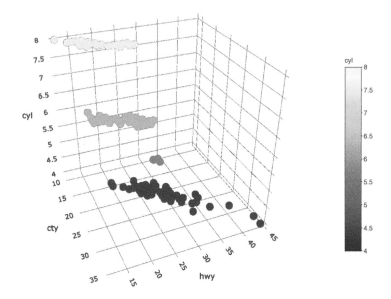

FIGURE 8.1: A 3D scatterplot.

8.2 Paths

To make a path in 3D, use add_paths() in the same way you would for a
2D path, but add a third variable z, as Figure 8.2 does.

```
plot_ly(mpg, x = ~cty, y = ~hwy, z = ~cyl) %>%
  add_paths(color = ~displ)
```

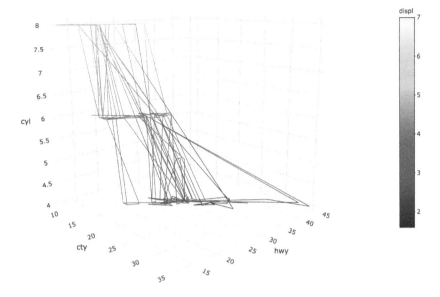

FIGURE 8.2: A path with color interpolation in 3D.

8.3 Lines

Figure 8.3 uses `add_lines()` instead of `add_paths()` to ensure the points are connected by the x axis instead of the row ordering.

```
plot_ly(mpg, x = ~cty, y = ~hwy, z = ~cyl) %>%
  add_lines(color = ~displ)
```

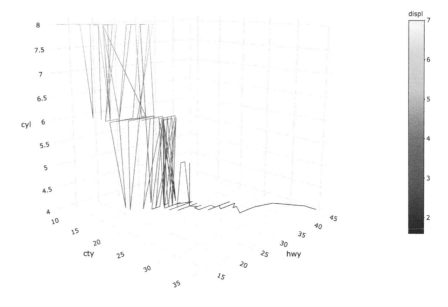

FIGURE 8.3: A line with color interpolation in 3D.

As with non-3D lines, you can make multiple lines by specifying a grouping variable.

```
plot_ly(mpg, x = ~cty, y = ~hwy, z = ~cyl) %>%
  group_by(cyl) %>%
  add_lines(color = ~displ)
```

FIGURE 8.4: Using `group_by()` to create multiple 3D lines.

8.4 Axes

For 3D plots, be aware that the axis objects are a part of the scene[1]
definition, which is part of the layout(). That is, if you wanted to set axis
titles (e.g., Figure 8.5), or something else specific to the axis definition,
the relation between axes (i.e., aspectratio[2]), or the default setting of
the camera (i.e., camera[3]); you would do so via the scence.

```
plot_ly(mpg, x = ~cty, y = ~hwy, z = ~cyl) %>%
  add_lines(color = ~displ) %>%
  layout(
    scene = list(
      xaxis = list(title = "MPG city"),
```

---

[1]https://plot.ly/r/reference/#layout-scene
[2]https://plot.ly/r/reference/#layout-scene-aspectratio
[3]https://plot.ly/r/reference/#layout-scene-camera

```
        yaxis = list(title = "MPG highway"),
        zaxis = list(title = "Number of cylinders")
    )
  )
```

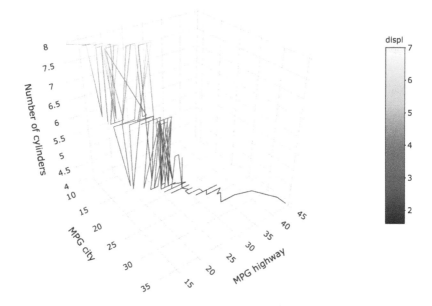

FIGURE 8.5: Setting axis titles on a 3D plot.

8.5   Surfaces

Creating 3D surfaces with add_surface() is a lot like creating heatmaps with add_heatmap(). In fact, you can even create 3D surfaces over categorical x/y (try changing add_heatmap() to add_surface() in Figure 7.3)! That being said, there should be a sensible ordering to the x/y axes in a surface plot since plotly.js interpolates z values. Usually the 3D surface is over a continuous region, as is done in Figure 8.6 to display the height of a volcano. If a numeric matrix is provided to z as in Figure 8.6, the x and y attributes do not have to be provided, but if they are, the length of x should match the number of columns in the matrix and y should match the number of rows.

```
x <- seq_len(nrow(volcano)) + 100
y <- seq_len(ncol(volcano)) + 500
plot_ly() %>% add_surface(x = ~x, y = ~y, z = ~volcano)
```

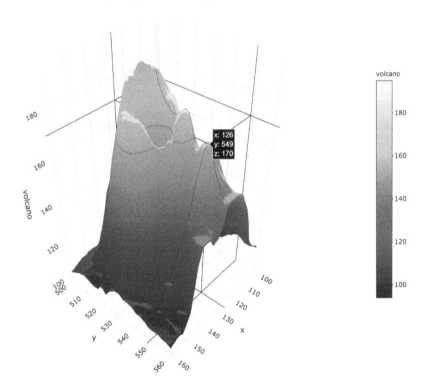

FIGURE 8.6: A 3D surface of volcano height.

Part II

Publishing views

9

Introduction

This portion of the book, Publishing Views, covers how to save **plotly** graphs as HTML, embed them within larger HTML documents, interactively edit (i.e., post-process) them, and export to static file formats — all of which can be useful tools for creating 'publication-quality' graphics. Static images can be created either from the command line (via the `orca()` function) or from the interactive graphic itself. The former is great if you need to export many images at once, and the latter is convenient if you need to export a manually edited version of the default view (e.g., Figure 12.1). All the R code in this chapter runs entirely locally using 100% free and open source software with no calls to external services.

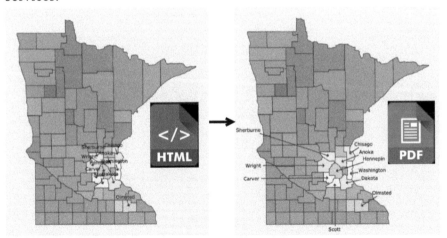

10

Saving and embedding HTML

Any widget made from any **htmlwidgets** package (e.g., **plotly**, **leaflet**, **DT**, etc.) can be saved as a standalone HTML file via the `htmlwidgets::saveWidget()` function. By default, it produces a completely self-contained HTML file, meaning that all the necessary JavaScript and CSS dependency files are bundled inside the HTML file. This makes it very easy to share a widget as a single HTML file. In this case, consider using the `partial_bundle()` function to reduce the size of the bundled files. By default, it automatically determines a reduced version of plotly.js that is sufficient for rendering your graphic. This can lead to a substantial reduction in the overall file size, especially if you're using basic chart types:

```
p <- plot_ly(x = 1:10, y = 1:10) %>% add_markers()
widget_file_size <- function(p) {
  d <- tempdir()
  withr::with_dir(d, htmlwidgets::saveWidget(p, "index.html"))
  f <- file.path(d, "index.html")
  mb <- round(file.info(f)$size / 1e6, 3)
  message("File is: ", mb," MB")
}
widget_file_size(p)
#> File is: 3.495 MB
widget_file_size(partial_bundle(p))
#> File is: 1.068 MB
```

If you want to embed numerous widgets in a larger HTML document (e.g., via HTML `<iframe>`s), *self-contained* HTML is not recommended. That's because, if you embed numerous self-contained widgets inside a larger document, your browser has to repeatedly parse the same dependencies over and over. Instead, if you save all the dependency files externally into a single directory, the browser will only have to parse those dependencies once, which can dramatically improve responsiveness. You can do this by setting `selfcontained = FALSE` and specifying a fixed `libdir` in `saveWidget()`. It's also worth noting that using `htmlwidgets::saveWidget()` with `selfcontained = FALSE` is essentially the same as using `htmltools::save_html()` which saves arbitrary HTML content to a file. The `htmltools::save_html()` function is useful for saving numerous htmlwidgets (e.g., Figure 13.12 or 13.13) and/or other custom HTML markup (e.g., Figure 22.1) in a single HTML page.

```r
library(htmlwidgets)
p <- plot_ly(x = rnorm(100))
saveWidget(p, "p1.html", selfcontained = F, libdir = "lib")
saveWidget(p, "p2.html", selfcontained = F, libdir = "lib")
```

In this case, if you wanted to share `"p1.html"` and/or `"p2.html"` with someone else, make sure to include the `libdir` folder, perhaps via a zip file:

```r
zip("p1.zip", c("p1.html", "lib"))
zip("p2.zip", c("p2.html", "lib"))
```

Embedding these HTML files via an HTML `<iframe>` is convenient not only for re-using a widget in various parent documents, but also for preventing any JavaScript and CSS in the parent document from negatively impacting how the widget renders. Rather than writing the HTML `<iframe>` tag directly, I recommend using `htmltools::tags$iframe()` – this will allow you to leverage **bookdown**'s figure captioning, numbering, and automatic snapshots for non-HTML output:

````r
```{r}
htmltools::tags$iframe(
  src = "p1.html",
  scrolling = "no",
  seamless = "seamless",
````
````

```
  frameBorder = "0"
)
```
```

A great tool that helps automate this sort of workflow with responsive iframes is the **widgetframe** package (Karambelkar, 2017). See the 'widgetframe and knitr' vignette for documentation of options for controling where, how, and if external dependencies are stored on the file system when using it inside a **knitr/rmarkdown** document.

```
browseVignettes("widgetframe")
```
```

11

Exporting static images

11.1 With code

Any **plotly** object can be saved as a static image via the `orca()` function. To use it, you'll need the `orca` command-line utility (CLI). This CLI can be installed via node.js, conda, or a standalone binary from `https://github.com/plotly/orca/releases`. Figure 11.1 demonstrates how `orca()` can generate a SVG (or PDF) that can then be imported into Adobe Illustrator for post-processing. Although it's a nice option to have, importing into Adobe Illustrator might not enable as nice of a workflow as using **plotly**'s native support for editable layout components in the browser, then exporting to SVG/PDF (as shown in Figure 12.1).

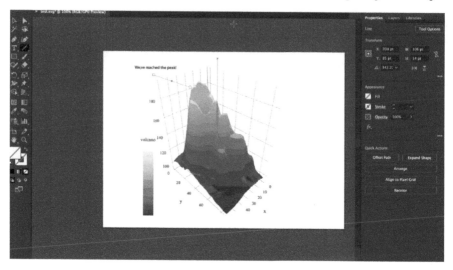

FIGURE 11.1: Using the `orca()` function to export a WebGL/Canvas based **plotly** graphic to a static PDF file. The resulting PDF file can then be imported into Adobe Illustrator for post-processing. For a video demonstration of the interactive, see https://bit.ly/plotly-orca. For the interactive, see https://plotly-r.com/interactives/orca.html

11.2   From a browser

Exporting an image from a browser is a nice option if you need to perform edits before exporting or if you'd like others to share your work. By default, the 'download plot' icon in the modebar will download to PNG and use the `height` and `width` of the plot, but these defaults can be altered via the plot's configuration, as done in Figure 11.2:

```
plot_ly() %>%
  config(
    toImageButtonOptions = list(
      format = "svg",
      filename = "myplot",
      width = 600,
```

```
    height = 700
  )
)
```

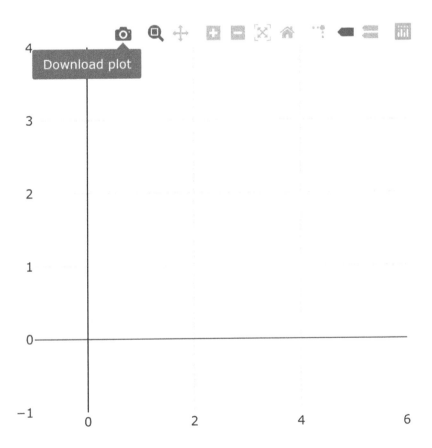

FIGURE 11.2: Specifying options for static image exporting via the modebar. Clicking on the 'download plot' icon should prompt your browser to download a static SVG file named 'myplot.svg'.

11.3   Sizing exports

It's worth noting that the height and width of a static image must be specified in pixels, which is intuitive for most file formats (PNG, JPG, SVG, etc.) but when exporting to PDF, you might want to specify the size in inches. If you multiply the DPI of your machine's display by the number of inches you want, you'll get the desired result. So, if you want a 8x11 PDF, and are on a typical 96 DPI display, you can do:

```
orca(plot_ly(), width = 8 * 96, height = 11 * 96)
```

On the other hand, if you're performing interactive editing and exporting, you may want to set a fixed size for the plot:

```
plot_ly(width = 8 * 96, height = 11 * 96) %>%
  config(toImageButtonOptions = list(format = "svg"))
```

12

Editing views for publishing

Numerous layout components of a **plotly** graph can be directly manipulated, including annotation text and placement (more on this in Section 17.2.3). In addition, the download (aka, toImage) button can be customized to export a static version to different file types including: SVG, PNG, JPG, and WebP. Since SVG can be easily converted to PDF, this effectively means we can edit a graph in a browser to perform touch-ups, then export to a high-quality PDF. At least currently, this workflow is recommended over first exporting to PDF (via `orca()`) then using Adobe Illustrator to manipulate the vectors, especially for adjusting the placement of annotations.

Figure 12.1 demonstrates this workflow on a choropleth map of estimated income in Minnesota by county where the top 10 counties by total income are labeled.[1]. For visuals like this, automated algorithms for placing the labels may not yield polished results, so it can be nice to have the option to adjust the placement manually. Although pressing 'download plot' exports a static version of the current state of the plot, there currently isn't an official way to save the state of these manual edits in the HTML version. You could, however, create a shiny app that listens to the `'plotly_relayout'` event to obtain the new annotation positions (see, for example, Figure 17.5) and translate that information into code.

```
library(dplyr)
library(sf)
library(purrr)
library(tidycensus)
```

---

[1]For perceptual reasons, you may want to consider using cartogram for this kind of map (see Section 4.2.2)

```r
library(USAboundaries)

# obtain geographical information for each county in MN
mn_sf <- us_counties(states = "MN")

# get income information for each county in MN
mn_income <- get_acs(
    geography = "county", variables = "B19013_001", state = "MN"
  ) %>%
  mutate(
    NAME = sub("County, Minnesota", "", NAME),
    county = reorder(NAME, estimate),
    color = scales::col_numeric("viridis", NULL)(estimate)
  )

# find center of each county (for placing annotations)
mn_center <- mn_sf %>%
  st_centroid() %>%
  mutate(
    x = map_dbl(geometry, 1),
    y = map_dbl(geometry, 2)
  )

# get top 10 counties by income with their x/y center location
top10labels <- mn_income %>%
  top_n(10, estimate) %>%
  left_join(mn_center, by = c("GEOID" = "geoid"))

# the map and top 10 county labels
map <- plot_ly() %>%
  add_sf(
    data = left_join(mn_sf, mn_income, by = c("geoid" = "GEOID")),
    color = ~I(color), split = ~NAME,
    stroke = I("black"), span = I(1), hoverinfo = "none"
  ) %>%
  add_annotations(
    data = select(top10labels, NAME, x, y),
```

```
    text = ~NAME, x = ~x, y = ~y
  )

# the dot-plot
bars <- ggplot(mn_income, aes(x = estimate, y = county)) +
  geom_errorbarh(aes(xmin = estimate - moe, xmax = estimate + moe)) +
  geom_point(aes(color = color), size = 2) +
  scale_color_identity()

# make manual edits in the browser, then click the
# 'toImage' button to export an SVG file
ggplotly(
  bars, dynamicTicks = TRUE, tooltip = "y",
  height = 8 * 96, width = 11 * 96
) %>%
  subplot(map, nrows = 1, widths = c(0.3, 0.7)) %>%
  layout(showlegend = FALSE) %>%
  config(
    edits = list(
      annotationPosition = TRUE,
      annotationTail = TRUE,
      annotationText = TRUE
    ),
    toImageButtonOptions = list(format = "svg")
  )
```

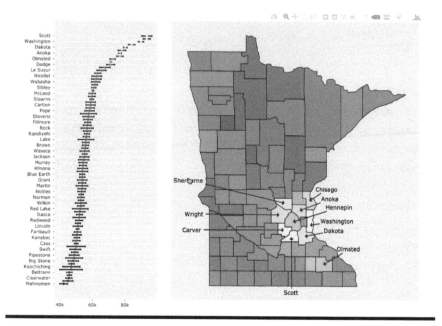

FIGURE 12.1: Estimated total income by county in Minnesota. The top ten counties are labeled with editable annotations. After manually adjusting the placement of county labels in a web browser, and zooming in on the top ten counties in the dot plot, the 'download plot' button is used to export to SVG. For a video demonstration of the interactive, see `https://bit.ly/edit-county-labels`. For the interactive, see `https://plotly-r.com/interactives/edit-county-labels.html`

After pressing the "download plot" button to export SVG, then the **rsvg** package can be used to convert the SVG to PDF (Ooms, 2018).

```r
# This is the directory that my browser places downloads...
# you may have to change this to your download directory
withr::with_dir(
  "~/Downloads/",
  rsvg::rsvg_pdf("newplot.svg", "mn.pdf")
)
```

Part III

Combining multiple views

13

Arranging views

One technique essential to high-dimensional data visualization is the ability to arrange multiple views. By arranging multiple low-dimensional graphics of the same (or similar) high-dimensional data, one can put local summaries and patterns into a global context. When arranging multiple **plotly** objects, you have some flexibility in terms of how you arrange them: you could use `subplot()` to merge multiple **plotly** objects into a single object (useful for synchronizing zoom/pan events across multiple axes), place them in separate HTML tags (Section 13.2), or embedded in a larger system for intelligently managing many views (Section 13.3).

Ideally, when displaying multiple related data views, they are linked through an underlying data source to foster comparisons and enable posing of data queries (Cook et al., 2007). Chapter 16.1 shows how to build upon these methods for arranging views to link them (client-side) as well.

13.1 Arranging plotly objects

The `subplot()` function provides a flexible interface for merging multiple **plotly** objects into a single object. It is more flexible than most trellis display frameworks (e.g., **ggplot2**'s `facet_wrap()`) as you don't have to condition on a value of common variable in each display (Becker et al., 1996). Its capabilities and interface are similar to the `grid.arrange()` function from the **gridExtra** package, which allows you to arrange multiple **grid** grobs in a single view, effectively providing a way to arrange (possibly unrelated) **ggplot2** and/or **lattice** plots in a single view (R Core Team, 2016; Auguie, 2016; Sarkar, 2008). Figure 13.1 shows

the most simple way to use subplot() which is to directly supply plotly
objects.

```
library(plotly)
p1 <- plot_ly(economics, x = ~date, y = ~unemploy) %>%
  add_lines(name = "unemploy")
p2 <- plot_ly(economics, x = ~date, y = ~uempmed) %>%
  add_lines(name = "uempmed")
subplot(p1, p2)
```

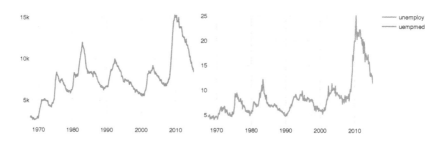

FIGURE 13.1: The most basic use of subplot() to merge multiple plotly
objects into a single plotly object.

Although subplot() accepts an arbitrary number of plot objects, passing
a *list* of plots can save typing and redundant code when dealing with
a large number of plots. Figure 13.2 shows one time series for each
variable in the economics dataset and shares the x-axis so that zoom/pan
events are synchronized across each series:

```
vars <- setdiff(names(economics), "date")
plots <- lapply(vars, function(var) {
  plot_ly(economics, x = ~date, y = as.formula(paste0("~", var))) %>%
    add_lines(name = var)
})
subplot(plots, nrows = length(plots), shareX = TRUE, titleX = FALSE)
```

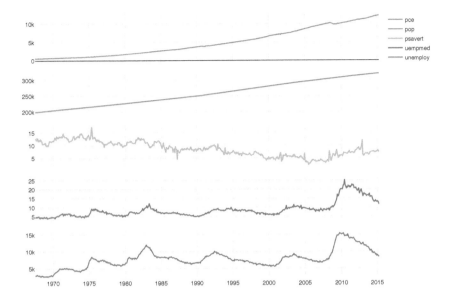

FIGURE 13.2: Five different economic variables on different y scales and a common x scale. Zoom and pan events in the x-direction are synchronized across plots.

Conceptually, subplot() provides a way to place a collection of plots into a table with a given number of rows and columns. The number of rows (and, by consequence, the number of columns) is specified via the nrows argument. By default, each row/column shares an equal proportion of the overall height/width, but as shown in Figure 13.3, the default can be changed via the heights and widths arguments.

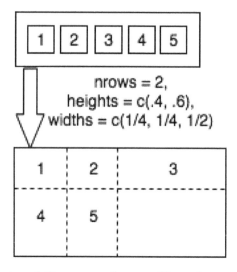

FIGURE 13.3: A visual diagram of controlling the `heights` of rows and `widths` of columns. In this particular example, there are five plots being placed in two rows and three columns.

This flexibility is quite useful for a number of visualizations, for example, as shown in Figure 13.4, a joint density plot is really a subplot of joint and marginal densities. The **heatmaply** package is a great example of leveraging `subplot()` in a similar way to create interactive dendrograms (Galili, 2016).

```
# draw random values from correlated bi-variate normal distribution
s <- matrix(c(1, 0.3, 0.3, 1), nrow = 2)
m <- mvtnorm::rmvnorm(1e5, sigma = s)
x <- m[, 1]
y <- m[, 2]
s <- subplot(
  plot_ly(x = x, color = I("black")),
  plotly_empty(),
  plot_ly(x = x, y = y, color = I("black")) %>%
    add_histogram2dcontour(colorscale = "Viridis"),
  plot_ly(y = y, color = I("black")),
  nrows = 2, heights = c(0.2, 0.8), widths = c(0.8, 0.2), margin = 0,
  shareX = TRUE, shareY = TRUE, titleX = FALSE, titleY = FALSE
```

```
)
layout(s, showlegend = FALSE)
```

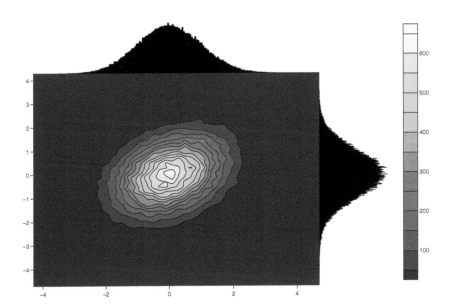

FIGURE 13.4: A joint density plot with synchronized axes.

13.1.1 Recursive subplots

The `subplot()` function returns a plotly object so it can be modified like any other plotly object. This effectively means that subplots work recursively (i.e., you can have subplots within subplots). This idea is useful when your desired layout doesn't conform to the table structure described in the previous section. In fact, you can think of a subplot of subplots like a spreadsheet with merged cells. Figure 13.5 gives a basic example where each row of the outermost subplot contains a different number of columns.

```
plotList <- function(nplots) {
  lapply(seq_len(nplots), function(x) plot_ly())
}
```

```
s1 <- subplot(plotList(6), nrows = 2, shareX = TRUE, shareY = TRUE)
s2 <- subplot(plotList(2), shareY = TRUE)
subplot(
  s1, s2, plot_ly(), nrows = 3,
  margin = 0.04, heights = c(0.6, 0.3, 0.1)
)
```

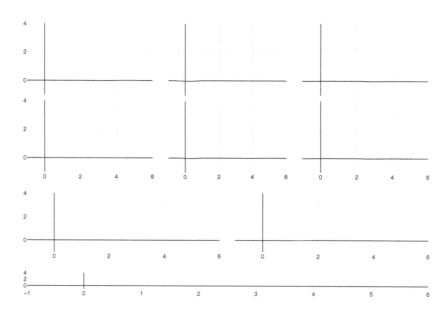

FIGURE 13.5: Recursive subplots.

The concept is particularly useful when you want plot(s) in a given row to have different widths from plot(s) in another row. Figure 13.6 uses this recursive behavior to place many bar charts in the first row, and a single choropleth in the second row.

```
# specify some map projection/options
g <- list(
  scope = 'usa',
  projection = list(type = 'albers usa'),
  lakecolor = toRGB('white')
```

```
)
# create a map of population density
density <- state.x77[, "Population"] / state.x77[, "Area"]
map <- plot_geo(
  z = ~density, text = state.name,
  locations = state.abb, locationmode = 'USA-states'
) %>%
  layout(geo = g)
# create a bunch of horizontal bar charts
vars <- colnames(state.x77)
barcharts <- lapply(vars, function(var) {
  plot_ly(x = state.x77[, var], y = state.name) %>%
    add_bars(orientation = "h", name = var) %>%
    layout(showlegend = FALSE, hovermode = "y",
           yaxis = list(showticklabels = FALSE))
})
subplot(barcharts, margin = 0.01) %>%
  subplot(map, nrows = 2, heights = c(0.3, 0.7), margin = 0.1) %>%
  layout(legend = list(y = 1)) %>%
  colorbar(y = 0.5)
```

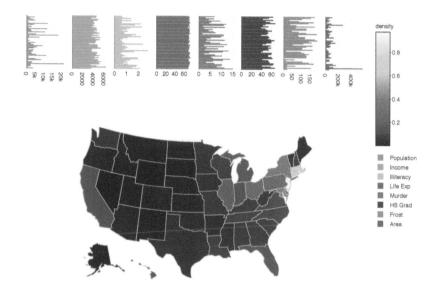

FIGURE 13.6: Multiple bar charts of U.S. statistics by state in a subplot with a choropleth of population density.

13.1.2 Other approaches and applications

Using `subplot()` directly is not the *only* way to create multiple views of a dataset with **plotly**. In some special cases, like scatterplot matrices and generalized pair plots, we can take advantage of some special methods designed specifically for these use cases.

13.1.2.1 Scatterplot matrices

The plotly.js library provides a trace specifically designed and optimized for scatterplot matrices (splom). To use it, provide numeric variables to the `dimensions` attribute of the `splom` trace type.

```
dims <- dplyr::select_if(iris, is.numeric)
dims <- purrr::map2(dims, names(dims), ~list(values=.x, label=.y))
plot_ly(
  type = "splom", dimensions = setNames(dims, NULL),
  showupperhalf = FALSE, diagonal = list(visible = FALSE)
)
```

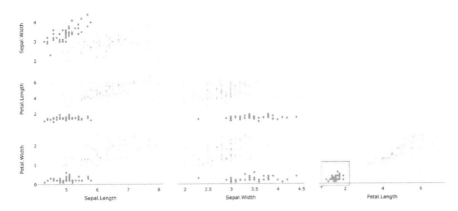

FIGURE 13.7: Linked brushing in a scatterplot matrix of the Iris dataset. For a video demonstration of the interactive, see `https://bit.ly/plotly-splom`. For the interactive, see `https://plotly-r.com/interactives/splom.html`

See `https://plot.ly/r/splom/` for more options related to the splom trace type.

13.1.2.2 Generalized pairs plot

The generalized pairs plot is an extension of the scatterplot matrix to support both discrete and numeric variables (Emerson et al., 2013). The `ggpairs()` function from the **GGally** package provides an interface for creating these plots via **ggplot2** (Schloerke et al., 2016). To implement `ggpairs()`, **GGally** introduces the notion of a matrix of **ggplot2** plot objects that it calls `ggmatrix()`. As Figure 13.8 shows, the `ggplotly()` function has a method for converting ggmatrix objects directly:

```
pm <- GGally::ggpairs(iris, aes(color = Species))
class(pm)
#> [1] "gg"  "ggmatrix"
ggplotly(pm)
```

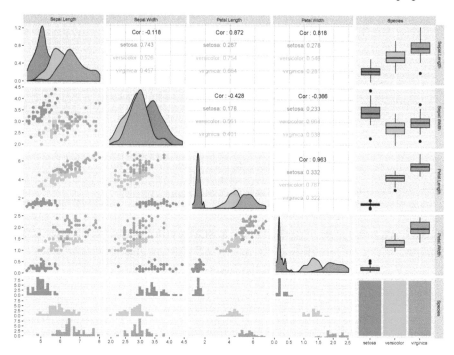

FIGURE 13.8: A generalized pairs plot made via the ggpairs() function from the **GGally** package.

As it turns out, **GGally** use ggmatrix() as a building block for other visualizations, like model diagnostic plots (ggnostic()). Sections 16.4.6 and 16.4.7 demonstrate how to leverage linked brushing in the ggplotly() versions of these plots.

13.1.2.3 Trellis displays with subplot()

It's true that **ggplot2**'s facet_wrap()/facet_grid() provides a simple way to create trellis displays, but for learning purposes, it can be helpful to learn how to implement a similar trellis display with plot_ly() and subplot(). Figure 13.9 demonstrates one approach, which leverages subplot()'s ability to reposition annotations and shapes. Specifically, the panel() function below, which defines the visualization method to be applied to each variable in the economics_long dataset, uses paper coordinates (i.e., graph coordinates on a normalized 0-1 scale) to place an annotation at the top-center of each panel as well as a rectangle shape behind the annotation. Note also the use of ysizemode =

'pixel' which gives the rectangle shape a fixed height (i.e., the rectangle height is always 16 pixels, regardless of the height of the trellis
display).

```r
library(dplyr)

panel <- . %>%
  plot_ly(x = ~date, y = ~value) %>%
  add_lines() %>%
  add_annotations(
    text = ~unique(variable),
    x = 0.5,
    y = 1,
    yref = "paper",
    xref = "paper",
    yanchor = "bottom",
    showarrow = FALSE,
    font = list(size = 15)
  ) %>%
  layout(
    showlegend = FALSE,
    shapes = list(
      type = "rect",
      x0 = 0,
      x1 = 1,
      xref = "paper",
      y0 = 0,
      y1 = 16,
      yanchor = 1,
      yref = "paper",
      ysizemode = "pixel",
      fillcolor = toRGB("gray80"),
      line = list(color = "transparent")
    )
  )
```

```
economics_long %>%
  group_by(variable) %>%
  do(p = panel(.)) %>%
  subplot(nrows = NROW(.), shareX = TRUE)
```

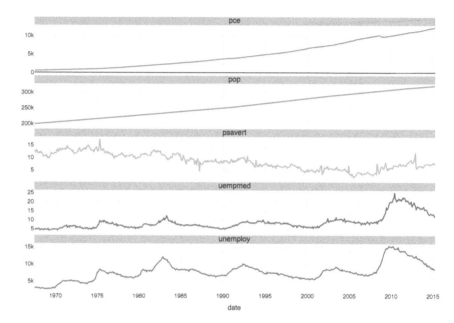

FIGURE 13.9: Creating a trellis display with `subplot()`.

13.1.2.4 ggplot2 subplots

It's possible to combine the convenience of **ggplot2**'s `facet_wrap()`/`facet_grid()` with the more flexible arrangement capabilities of `subplot()`. Figure 13.10 does this to show two different views of the `economics_long` data: the left-hand column displays each variable along time, while the right-hand column shows violin plots of each variable. For the implementation, each column is created through `ggplot2::facet_wrap()`, but then the trellis displays are combined with `subplot()`. In this case, **ggplot2** objects are passed directly to `subplot()`, but you can also use `ggplotly()` for finer control over the conversion of **ggplot2** to **plotly** (see also Chapter 33) before supplying that result to `subplot()`.

```r
gg1 <- ggplot(economics_long, aes(date, value)) + geom_line() +
  facet_wrap(~variable, scales = "free_y", ncol = 1)
gg2 <- ggplot(economics_long, aes(factor(1), value)) +
  geom_violin() +
  facet_wrap(~variable, scales = "free_y", ncol = 1) +
  theme(axis.text = element_blank(), axis.ticks = element_blank())
subplot(gg1, gg2)
```

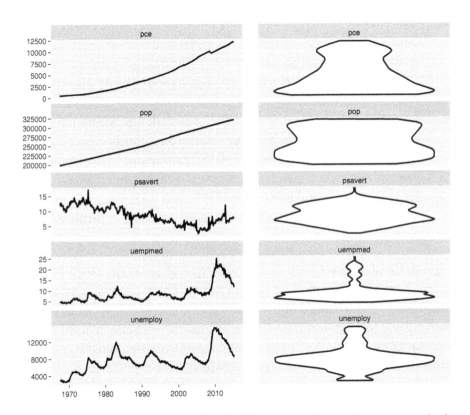

FIGURE 13.10: Arranging multiple-faceted ggplot2 plots into a plotly subplot.

13.2   Arranging htmlwidgets

Since **plotly** objects are also **htmlwidgets**, any method that works for arranging **htmlwidgets** also works for **plotly** objects. Moreover, since **htmlwidgets** are also **htmltools** tags, any method that works for arranging **htmltools** tags also works for **htmlwidgets**. Here are three common ways to arrange components (e.g., **htmlwidgets**, **htmltools** tags, etc.) in a single webpage:

1. **flexdashboard**: An R package for arranging components into an opinionated dashboard layout. This package is essentially a special **rmarkdown** template that uses a simple markup syntax to define the layout.
2. Bootstrap's grid layout: Both the **crosstalk** and **shiny** packages provide ways to arrange numerous components via Bootstrap's (a popular HTML/CSS framework) grid layout system[1].
3. CSS flexbox: If you know some HTML and CSS, you can leverage CSS flexbox[2] to arrange components via the **htmltools** package.

Although **flexdashboard** is a really excellent way to arrange web-based content generated from R, it can payoff to know the other two approaches as their arrangement techniques are agnostic to an **rmarkdown** output format. In other words, approaches 2-3 can be used with any **rmarkdown** template[3] or really *any* framework for website generation. Although Bootstrap grid layout system (2) is expressive and intuitive, using it in a larger website that also uses a different HTML/CSS framework (e.g., Bulma, Skeleton, etc.) can cause issues. In that case,

---

[1] https://getbootstrap.com/docs/4.1/layout/grid/

[2] https://css-tricks.com/snippets/css/a-guide-to-flexbox/

[3] Although HTML cannot possibly render in a PDF or Word document, **knitr** can automatically detect a non-HTML output format and embed a static image of the htmlwidget via the **webshot** package (Chang, 2016).

CSS flexbox (3) is a lightweight (i.e., no external CSS/JS dependencies) alternative that is less likely to introduce undesirable side effects.

13.2.1 flexdashboard

Figure 13.11 provides an example of embedding `ggplotly()` inside **flexdashboard** (Allaire, 2016). Since **flexdashboard** is an **rmarkdown** template, it automatically comes with many things that make **rmarkdown** great: ability to produce standalone HTML, integration with other languages, and thoughtful integration with RStudio products like Connect. There are many other things to like about **flexdashboard**, including lots of easy-to-use theming options, multiple pages, storyboards, and even **shiny** integration. Explaining how the **flexdashboard** package actually works is beyond the scope of this book, but you can visit the website for documentation and more examples `https://rmarkdown.rstudio.com/flexdashboard/`.

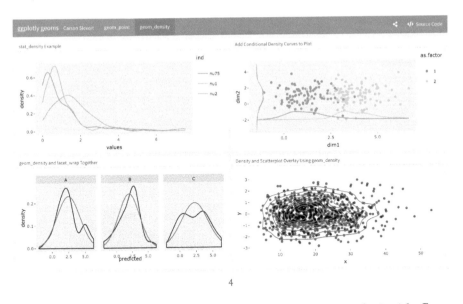

FIGURE 13.11: An example of embedding `ggplotly()` graphs inside **flexdashboard**. See here for the interactive dashboard `https://plotly-r.com/flexdashboard.html`

13.2.2   Bootstrap grid layout

If you're already familiar with **shiny**, you may already be familiar with functions like `fluidPage()`, `fluidRow()`, and `column()`. These R functions provide an interface from R to Bootstrap's grid layout system. That layout system is based on the notion of rows and columns where each row spans a width of 12 columns. Figure 13.12 demonstrates how one can use these functions to produce a standalone HTML page with three **plotly** graphs — with the first plot in the first row spanning the full width and the other 2 plots in the second row of equal width. To learn more about this `fluidPage()` approach to layouts, see `https:` `//shiny.rstudio.com/articles/layout-guide.html`.

```
library(shiny)
p <- plot_ly(x = rnorm(100))
fluidPage(
  fluidRow(p),
  fluidRow(
    column(6, p), column(6, p)
  )
)
```

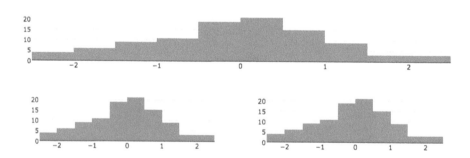

FIGURE 13.12: Arranging multiple **htmlwidgets** with `fluidPage()` from the **shiny** package.

It's also worth noting another, somewhat similar, yet more succinct, interface to grid's layout system provided by the `bscols()` function from the **crosstalk** package. You can think of it in a similar way to `fluidRow()`, but instead of defining `column()` width for each component individually, you can specify the width of several components at once through the `widths` argument. Also, importantly, this function works recursively; it returns a collection of **htmltools** tags and accepts them as input as well. The code below produces the same result as above, but is a much more succinct way of doing so.

```
bscols(p, bscols(p, p), widths = 12)
```

Bootstrap is much more than just its grid layout system, so beware; using either of these approaches will impose Bootstrap's styling rules on other content in your webpage. If you are using another Cascading Style Sheet (CSS) framework for styling or just want to reduce the size of dependencies in your webpage, consider working with CSS flexbox instead of Bootstrap.

13.2.3   CSS flexbox

Cascading Style Sheet (CSS) flexbox is a relatively new CSS feature that most modern web browsers natively support.[5] It aims to provide a general system for distributing space among multiple components in a container. Instead of covering this entire system, we'll cover its basic functionality, which is fairly similar to Bootstrap's grid layout system.

Creating a flexbox requires a flexbox container; in HTML speak, that means a `<div>` tag with a CSS style property of `display: flex`. By default, in this display setting, all the components inside that container will try fitting in a single row. To allow 'overflowing' components the freedom to 'wrap' into new row(s), set the CSS property of `flex-wrap: wrap` in the parent container. Another useful CSS property to know about for the 'parent' container is `justify-content:` in the case of Figure 13.13, I'm using it to horizontally `center` the components. Moreover, since I've imposed a width of 40% for the first two plots, the net effect is that we have 2 plots in the first two (spanning 80% of the page width), then the third plot wraps onto a new line.

---

[5]For a full reference of which browsers/versions support flexbox, see `https:// caniuse.com/#feat=flexbox`.

```
library(htmltools)
p <- plot_ly(x = rnorm(100))
# NOTE: you don't need browsable() in rmarkdown,
# but you do at the R prompt
browsable(div(
  style = "display: flex; flex-wrap: wrap; justify-content: center",
  div(p, style = "width: 40%; border: solid;"),
  div(p, style = "width: 40%; border: solid;"),
  div(p, style = "width: 100%; border: solid;")
))
```

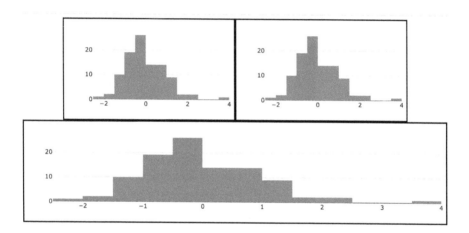

FIGURE 13.13: Arranging multiple **htmlwidgets** with CSS flexbox.

From the code example in Figure 13.13, you might notice that display: flex; flex-wrap: wrap is quite similar to Bootstrap grid layout system. The main difference is that, instead of specifying widths in terms of 12 columns, you have more flexibility with how to size things, as well as how you handle extra space. Here, in Figure 13.13 I've used widths that are relative to the page width, but you could also use fixed widths (using fixed widths, however, is generally frowned upon). For those who would like to learn about more details about CSS flexbox, see https://css-tricks.com/snippets/css/a-guide-to-flexbox/.

13.3   Arranging many views

As we've already seen in Figures 2.10, 16.14, and 13.9, the trellis (aka small multiple) display is an effective way to see how a conditional distribution behaves under different conditions. In other words, the trellis display helps us understand how patterns or structure in the data changes across groups. However, trellis displays do have a limitation: they don't scale very well to a large number of groups.

Before trellis displays were formally introduced, Tukey, J. and Tukey P. (1985) proposed a solution to the problem of scatterplots not being able to scale to a large number of variables (i.e., it's time consuming to visualize 1000 scatterplots!). The proposed solution involved using quantitative measurements of various scatterplot characteristics (e.g., correlation, clumpiness, etc.) to help summarise and guide attention towards 'interesting' scatterplots. This idea, coined scagnostics (short for scatterplot diagnostics), has since been made explicit, and many other similar applications have been explored, even techniques for time-series (Wilkinson et al., 2005; Wilkinson and Wills, 2008; Dang and Wilkinson, 2012). The idea of associating quantitative measures with a graphical display of data can be generalized to include more than just scatterplots, and in this more general case, these measures are sometimes referred to as cognostics.

In addition to being useful for navigating exploration of many variables, cognostics can also be useful for exploring many subsets of data. This idea has inspired work on more general divide and recombine technique(s) for working with navigating through many statistical artifacts (Cleveland and Hafen, 2014; Guha et al., 2012), including visualizations (Hafen et al., 2013). The **trelliscope** package provides a system for computing arbitrary cognostics on each panel of a trellis display as well as an interactive graphical user interface for defining (and navigating through) interesting panels based on those cognostics (Hafen, 2016). This system also allows users to define the graphical method for displaying each panel, so **plotly** graphs can easily be embedded. The **trelliscope** package is currently built upon **shiny**, but as Figure 13.14 demonstrates, the **trelliscopejs** package provides lower-level tools that allow one to create trelliscope displays without **shiny** (Hafen and Schloerke, 2018).

As the video behind Figure 13.14 demonstrates, **trelliscopejs** provides two very powerful interactive techniques for surfacing 'interesting' panels: sorting and filtering. In this toy example, each panel represents a different country, and the life expectancy is plotted as a function of time. By default, **trelliscopejs** sorts panels by group alphabetically, which is why, on page load we see the first 12 countries (Afghanistan, Albania, Algeria, etc.). By opening the sort menu, we can pick and sort by any cognostic for any variable in the dataset. If no cognostics are supplied (as it the case here), some sensible ones are computed and supplied for us (e.g., mean, median, var, max, min). In this case, since we are primarily interested in life expectancy, we sort by life expectancy. This simple task allows us to quickly see the countries with the best and worst average life expectancy, as well as how it has evolved over time. By combining sort with filter, we can surface countries that perform well/poorly under certain conditions. For example, Cuba, Uruguay, Taiwan have great life expectancy considering their GDP per capita. Also, within the Americas, Haiti, Bolivia, and Guatemala have the poorest life expectancy.

```
library(trelliscopejs)
data(gapminder, package = "gapminder")

qplot(year, lifeExp, data = gapminder) +
  xlim(1948, 2011) + ylim(10, 95) + theme_bw() +
  facet_trelliscope(~ country + continent,
    nrow = 2, ncol = 6, width = 300,
    as_plotly = TRUE,
    plotly_args = list(dynamicTicks = T),
    plotly_cfg = list(displayModeBar = F)
  )
```

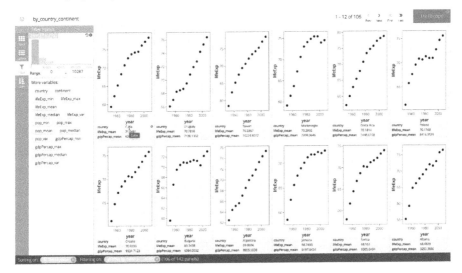

FIGURE 13.14: Using **trelliscopejs** to surface high-dimensional insights related to life expectancy and GDP per capita in various countries. For a video demonstration of the interactive, see `https://bit.ly/plotly-trelliscope`. For the interactive, see `https://plotly-r.com/interactives/trelliscope`

14

Animating views

14.1 Animation API

Both `plot_ly()` and `ggplotly()` support key frame[1] animations through the `frame` argument/aesthetic. They also support an `ids` argument/aesthetic to ensure smooth transitions between objects with the same id (which helps facilitate object constancy[2]). Figure 14.1 recreates the famous gapminder animation of the evolution in the relationship between GDP per capita and life expectancy evolved over time (Bryan, 2015). The data is recorded on a yearly basis, so the year is assigned to `frame`, and each point in the scatterplot represents a country, so the country is assigned to `ids`, ensuring a smooth transition from year to year for a given country.

```
data(gapminder, package = "gapminder")
gg <- ggplot(gapminder, aes(gdpPercap, lifeExp, color = continent)) +
  geom_point(aes(size = pop, frame = year, ids = country)) +
  scale_x_log10()
ggplotly(gg)
```

---

[1] https://en.wikipedia.org/wiki/Key_frame
[2] https://bost.ocks.org/mike/constancy/

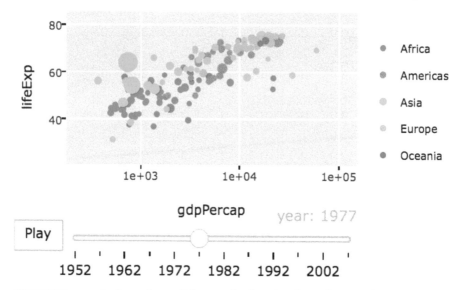

FIGURE 14.1: Animation of the evolution in the relationship between GDP per capita and life expectancy in numerous countries.

As long as a `frame` variable is provided, an animation is produced with play/pause button(s) and a slider component for controlling the animation. These components can be removed or customized via the `animation_button()` and `animation_slider()` functions. Moreover, various animation options, like the amount of time between frames, the smooth transition duration, and the type of transition easing may be altered via the `animation_opts()` function. Figure 14.2 shows the same data as Figure 14.1, but doubles the amount of time between frames, uses linear transition easing, places the animation buttons closer to the slider, and modifies the default `currentvalue.prefix` settings for the slider.

```
base <- gapminder %>%
  plot_ly(x = ~gdpPercap, y = ~lifeExp, size = ~pop,
          text = ~country, hoverinfo = "text") %>%
  layout(xaxis = list(type = "log"))

base %>%
  add_markers(color = ~continent, frame = ~year, ids = ~country) %>%
```

```
animation_opts(1000, easing = "elastic", redraw = FALSE) %>%
animation_button(
  x = 1, xanchor = "right", y = 0, yanchor = "bottom"
) %>%
animation_slider(
  currentvalue = list(prefix = "YEAR ", font = list(color="red"))
)
```

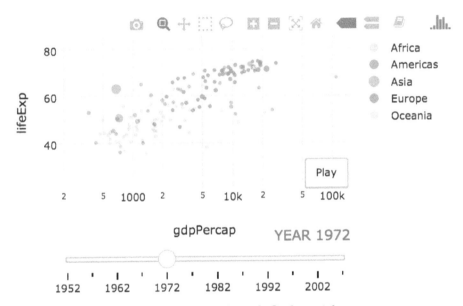

FIGURE 14.2: Modifying animation defaults with `animation_opts()`, `animation_button()`, and `animation_slider()`.

If `frame` is a numeric variable (or a character string), frames are always ordered in increasing (alphabetical) order; but for factors, the ordering reflects the ordering of the levels. Consequently, factors provide the most control over the ordering of frames. In Figure 14.3, the continents (i.e., frames) are ordered according to their average life expectancy across countries within the continent. Furthermore, since there is no meaningful relationship between objects in different frames of Figure 14.3, the smooth transition duration is set to 0. This helps avoid any confusion that there is a meaningful connection between the smooth transitions. Note that these options control both animations triggered by the play button or via the slider.

```
meanLife <- with(gapminder, tapply(lifeExp, INDEX = continent, mean))
gapminder$continent <- factor(
  gapminder$continent, levels = names(sort(meanLife))
)
```

```
base %>%
  add_markers(data = gapminder, frame = ~continent) %>%
  hide_legend() %>%
  animation_opts(frame = 1000, transition = 0, redraw = FALSE)
```

FIGURE 14.3: Animation of GDP per capita versus life expectancy by continent. The ordering of the continents goes from lowest average (across countries) life expectancy to highest.

Both the `frame` and `ids` attributes operate on the trace level, meaning that we can target specific layers of the graph to be animated. One obvious use case for this is to provide a background which displays every possible frame (which is not animated) and overlay the animated frames onto that background. Figure 14.4 shows the same information as Figure 14.2, but layers animated frames on top of a background of all the frames. As a result, it is easier to put a specific year into a global context.

```
base %>%
  add_markers(
    color = ~continent, showlegend = F,
    alpha = 0.2, alpha_stroke = 0.2
  ) %>%
  add_markers(color = ~continent, frame = ~year, ids = ~country) %>%
  animation_opts(1000, redraw = FALSE)
```

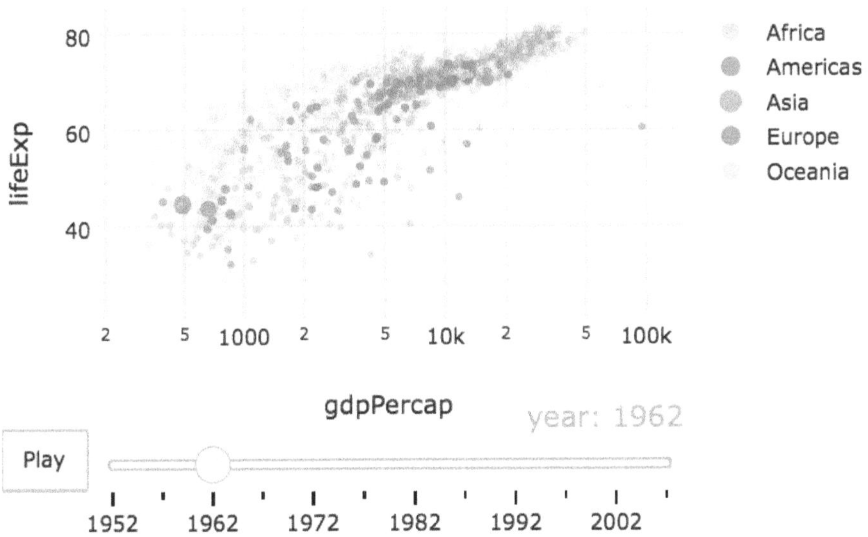

FIGURE 14.4: Overlaying animated frames on top of a background of all possible frames.

14.2 Animation support

At the time of writing, the scatter plotly.js trace type is really the only trace type with full support for animation. That means, we need to get a little imaginative to animate certain things, like a population pyramid chart (essentially a bar chart) using `add_segments()` (a scatter-based layer) instead of `add_bars()` (a non-scatter layer). Figure 14.5 shows projections for male and female population by age from 2018 to 2050 using data obtained via the **idbr** package (Walker, 2018).

```r
library(idbr)
library(dplyr)

us <- bind_rows(
  idb1(
    country = "US",
    year = 2018:2050,
    variables = c("AGE", "NAME", "POP"),
    sex = "male"
  ),
  idb1(
    country = "US",
    year = 2018:2050,
    variables = c("AGE", "NAME", "POP"),
    sex = "female"
  )
)

us <- us %>%
  mutate(
    POP = if_else(SEX == 1, POP, -POP),
    SEX = if_else(SEX == 1, "Male", "Female")
  )

plot_ly(us, size = I(5), alpha  = 0.5) %>%
  add_segments(
    x = ~POP, xend = 0,
    y = ~AGE, yend = ~AGE,
    frame = ~time,
```

```
    color = ~factor(SEX)
)
```

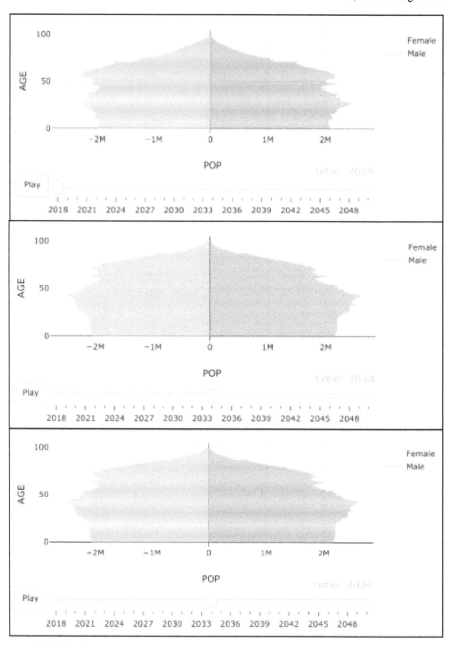

FIGURE 14.5: U.S. population projections by age and gender from 2018 to 2050. This population pyramid is implemented with thick line segments to give the appearance of bars. The still image shows three frames of the animation. For a video of the animation, see `https://bit.ly/profile-pyramid`. For the interactive, see `https://plotly-r.com/interactives/profile-pyramid.html`

Although population pyramids are quite popular, they aren't necessarily the best way to visualize this information, especially if the goal is to compare the population profiles over time. It's much easier to compare them along a common scale, as done in Figure 14.6. Note that, when animating lines in this fashion, it can help to set line.simplify[3] to FALSE so that the number of points along the path is left unaffected.

```
plot_ly(us, alpha  = 0.5) %>%
  add_lines(
    x = ~AGE, y = ~abs(POP),
    frame = ~time,
    color = ~factor(SEX),
    line = list(simplify = FALSE)
  ) %>%
  layout(yaxis = list(title = "US population"))
```

---

[3]https://plot.ly/r/reference/#scatter-line-simplify

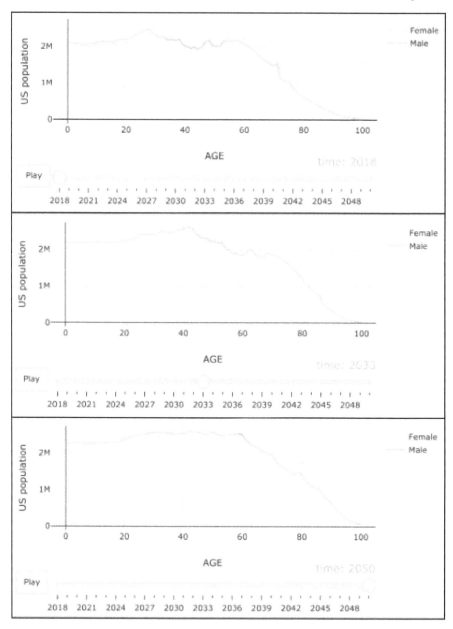

FIGURE 14.6: Visualizing the same information in Figure 14.5 using lines rather than segments. The still image shows three frames of the animation. For a video of the animation, see `https://bit.ly/profile-lines`. For the interactive, see `https://plotly-r.com/interactives/profile-lines.html`

Part IV

Linking multiple views

15

Introduction

Linking of multiple data views offers a powerful approach to visualization as well as communication of structure in high-dimensional data. In particular, linking of multiple 1-2 dimensional statistical graphics can often lead to insight that a single view could not possibly reveal. For decades, statisticians and computer scientists have been using and authoring systems for multiple linked views, many of which can be found in ASA's video library[1]. Some noteworthy videos include focussing and linking[2], missing values[3], and exploring Tour De France data[4] (Swayne et al., 1998; Theus and Urbanek, 2008).

These early systems were incredibly sophisticated, but the interactive graphics they produce are not easily shared, replicated, or incorporated in a larger document. Web technologies offer the infrastructure to address these issues, which is a big reason why many modern interactive graphics systems are now web based. When talking about interactive *web-based* graphics, it's important to recognize the difference between a web application and a purely client-side webpage, especially when it comes to saving, sharing, and hosting the result.

A web application relies on a client-server relationship where the client's (i.e., end user) web browser requests content from a remote server. This model is necessary whenever the webpage needs to execute computer code that is not natively supported by the client's web browser. As Chapter 17 details, the flexibility that a web application framework, like **shiny**, offers is an incredibly productive and powerful way to link multiple data views; but when it comes to distributing a

---

[1]http://stat-graphics.org/movies

[2]http://stat-graphics.org/movies/focussing-linking.html

[3]http://stat-graphics.org/movies/missing-data.html

[4]http://stat-graphics.org/movies/tour-de-france.html

web application, it introduces a lot of complexity and computational infrastructure that may or may not be necessary.

Figure 15.1 is a basic illustration of the difference between a web application and a purely client-side webpage. Thanks to `JavaScript` and `HTML5`, purely client-side webpages can still be dynamic without any software dependencies besides a modern web browser. In fact, Section 16.1 outlines **plotly**'s graphical querying framework for linking multiple plots entirely client-side, which makes the result very easy to distribute (see Chapter 10). There are, of course, many useful examples of linked and dynamic views that cannot be easily expressed as a database query, but a surprising amount actually can, and the remainder can likely be quickly implemented as a **shiny** web application.

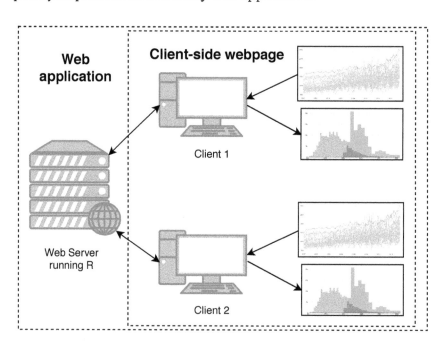

FIGURE 15.1: A diagram of the graphical querying framework underlying Figure 16.6.

The graphical querying framework implemented by **plotly** is inspired by Buja et al. (1991), where direct manipulation of graphical elements in multiple linked plots is used to perform database queries and visually reveal high-dimensional structure in real-time. Cook et al. (2007) goes on to argue this framework is preferable to posing database queries dynamically via menus, as described by Ahlberg et al. (1991), and goes on to state that, "Multiple linked views are the optimal framework for posing queries about data". The next section shows you how to implement similar graphical queries in a standalone webpage using R code.

16

Client-side linking

16.1 Graphical queries

This section focuses on a particular approach to linking views known as graphical (database) queries using the R package **plotly**. With **plotly**, one can write R code to pose graphical queries that operate entirely client-side in a web browser (i.e., no special web server or callback to R is required). In addition to teaching you how to pose queries with the `highlight_key()` function, this section shows you how to control queries that are triggered and visually rendered via the `highlight()` function.

Figure 16.1 shows a scatterplot of the relationship between weight and miles per gallon of 32 cars. It also uses `highlight_key()` to assign the number of cylinders to each point so that when a particular point is 'queried', all points with the same number of cylinders are highlighted (the number of cylinders is displayed with text just for demonstration purposes). By default, a mouse click triggers a query, and a double-click clears the query, but both of these events can be customized through the `highlight()` function. By typing `help(highlight)` in your R console, you can learn more about what events are supported for turning graphical queries on and off.

```
library(plotly)
mtcars %>%
  highlight_key(~cyl) %>%
  plot_ly(
    x = ~wt, y = ~mpg, text = ~cyl, mode = "markers+text",
    textposition = "top", hoverinfo = "x+y"
```

```
) %>%
highlight(on = "plotly_hover", off = "plotly_doubleclick")
```

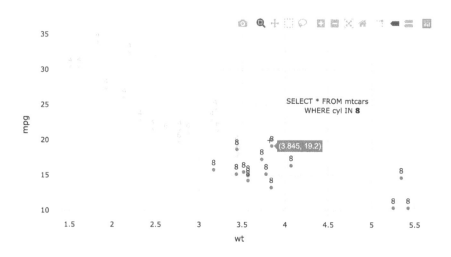

FIGURE 16.1: A visual depiction of how `highlight_key()` attaches meta-data to graphical elements to enable graphical database queries. Each point represents a different car and the number of cylinders (`cyl`) is assigned as metadata so that when a particular point is queried, all points with the same number of cylinders are highlighted. For a video demonstration of the interactive, see `https://bit.ly/Link-intro`. For the interactive, see `https://plotly-r.com/interactives/link-intro.html`

Generally speaking, `highlight_key()` assigns data values to graphical marks so that when graphical mark(s) are *directly manipulated* through the `on` event, it uses the corresponding data values (call it `$SELECTION_VALUE`) to perform an SQL query of the following form.

```
SELECT * FROM mtcars WHERE cyl IN $SELECTION_VALUE
```

For a more useful example, let's use graphical querying to pose interactive queries of the `txhousing` dataset. This data contains monthly housing sales in Texan cities acquired from the TAMU real estate center[1] and made available via the **ggplot2** package. Figure 16.2 shows the

---

[1]`http://recenter.tamu.edu/`

median house price in each city over time which produces a rather busy (spaghetti) plot. To help combat the overplotting, we could add the ability to click a particular data point along a line to highlight that particular city. This interactive ability is enabled by simply using highlight_key() to declare that the city variable be used as the querying criteria within the graphical querying framework.

One subtlety to be aware of in terms of what makes Figure 16.2 possible is that every point along a line may have a different data value assigned to it. In this case, since the city column is used as both the visual grouping *and* querying variable, we effectively get the ability to highlight a group by clicking on any point along that line. Section 16.4.1 has examples of using different grouping and querying variables to query multiple related groups of visual geometries at once, which can be a powerful technique.[2]

```r
# load the `txhousing` dataset
data(txhousing, package = "ggplot2")

# declare `city` as the SQL 'query by' column
tx <- highlight_key(txhousing, ~city)

# initiate a plotly object
base <- plot_ly(tx, color = I("black")) %>%
  group_by(city)

# create a time series of median house price
base %>%
  group_by(city) %>%
  add_lines(x = ~date, y = ~median)
```

---

[2]This sort of idea relates closely to the notion of generalized selections as described in Heer et al. (2008).

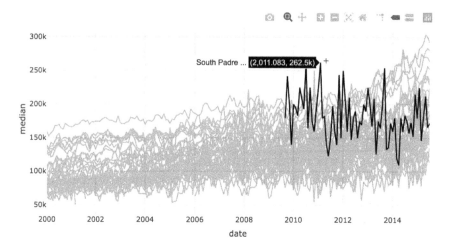

FIGURE 16.2: Graphical query of housing prices in various Texan cities. The query in this particular example must be triggered through clicking directly on a time series. For a video demonstration of the interactive, see `https://bit.ly/txmissing`. For the interactive, see `https://plotly-r.com/interactives/txmissing.html`

Querying a city via direct manipulation is somewhat helpful for focussing on a particular time series, but it's not so helpful for querying a city by name and/or comparing multiple cities at once. As it turns out, **plotly** makes it easy to add a selectize.js powered dropdown widget for querying by name (aka indirect manipulation) by setting `selectize` = `TRUE`.[3] When it comes to comparing multiple cities, we want to be able to both retain previous selections (`persistent` = `TRUE`) as well as control the highlighting color (`dynamic` = `TRUE`). This video explains how to use these features in Figure 16.3 to compare pricing across different cities.

```
highlight(
  time_series,
  on = "plotly_click",
```

---

[3]The title that appears in the dropdown can be controlled via the `group` argument in the `highlight_key()` function. The primary purpose of the `group` argument is to isolate one group of linked plots from others.

```
  selectize = TRUE,
  dynamic = TRUE,
  persistent = TRUE
)
```

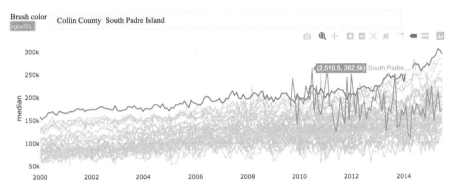

FIGURE 16.3: Using a selectize dropdown widget to search for cities by name and comparing multiple cities through persistent selection with a dynamic highlighting color. For a visual and audio explanation, see `https://bit.ly/txmissing-modes`.

By querying a few different cities in Figure 16.3, one obvious thing we can learn is that not every city has complete pricing information (e.g., South Padre Island, San Marcos, etc.). To learn more about what cities are missing information as well as how that missingness is structured, Figure 16.4 links a view of the raw time series to a dot-plot of the corresponding number of missing values per city. In addition to making it easy to see how cities rank in terms of missing house prices, it also provides a way to query the corresponding time series (i.e., reveal the structure of those missing values) by brushing cities in the dot-plot. This general pattern of linking aggregated views of the data to more detailed views fits the famous and practical information visualization advice from Shneiderman (1996): "Overview first, zoom and filter, then details on demand".

```
# remember, `base` is a plotly object, but we can use dplyr verbs to
```

```
# manipulate the input data
# (`txhousing` with `city` as a grouping and querying variable)
dot_plot <- base %>%
  summarise(miss = sum(is.na(median))) %>%
  filter(miss > 0) %>%
  add_markers(
    x = ~miss,
    y = ~forcats::fct_reorder(city, miss),
    hoverinfo = "x+y"
  ) %>%
  layout(
    xaxis = list(title = "Number of months missing"),
    yaxis = list(title = "")
  )

subplot(dot_plot, time_series, widths = c(.2, .8), titleX = TRUE) %>%
  layout(showlegend = FALSE) %>%
  highlight(on = "plotly_selected", dynamic = TRUE, selectize = TRUE)
```

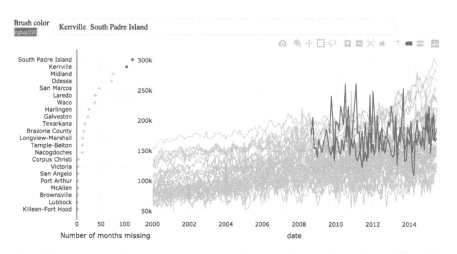

FIGURE 16.4: Linking a dot-plot of the number of missing housing prices with the raw time series. By brushing markers on the dot-plot, their raw time series is highlighted on the right-hand side.

How does **plotly** know to highlight the time series when markers in the dot-plot are selected? The answer lies in what data values are embedded in the graphical markers via `highlight_key()`. When 'South Padre Island' is selected, as in Figure 16.5, it seems as though the logic says to simply change the color of any graphical elements that match that value, but the logic behind **plotly**'s graphical queries is a bit more subtle and powerful. Another, more accurate, framing of the logic is to first imagine a linked database query being performed behind the scenes (as in Figure 16.5). When 'South Padre Island' is selected, it first filters the aggregated dot-plot data down to just that one row, then it filters down the raw time-series data down to every row with 'South Padre Island' as a city. The drawing logic will then call `Plotly.addTrace()`[4] with the newly filtered data which adds a new graphical layer representing the selection, allowing us to have fine-tuned control over the visual encoding of the data query.

---

[4] `https://plot.ly/javascript/plotlyjs-function-reference/#plotlyaddtraces`

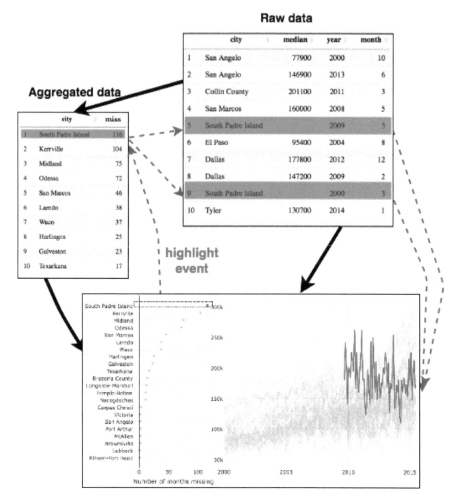

FIGURE 16.5: A diagram of the graphical querying framework underlying Figure 16.4.

The biggest advantage of drawing an entirely new graphical layer with the filtered data is that it becomes easy to leverage statistical trace types[5] for producing summaries that are conditional on the query. Figure 16.6 leverages this functionality to dynamically produce probability densities of house price in response to a query event. Section 16.4.2 has more examples of leveraging statistical trace types with graphical queries.

---

[5]https://plot.ly/r/statistical-charts/

```
hist <- add_histogram(
  base,
  x = ~median,
  histnorm = "probability density"
)
subplot(time_series, hist, nrows = 2) %>%
  layout(barmode = "overlay", showlegend = FALSE) %>%
  highlight(
    dynamic = TRUE,
    selectize = TRUE,
    selected = attrs_selected(opacity = 0.3)
  )
```

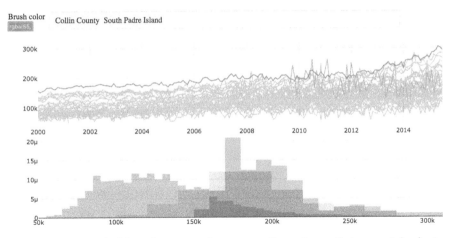

FIGURE 16.6: Linking house prices as a function of time with their probability density estimates.

Another neat consequence of drawing a completely new layer is that we can control the plotly.js attributes in that layer through the selected argument of the highlight() function. In Figure 16.6 we use it to ensure the new highlighting layer has some transparency to more easily compare the city specific distribution to the overall distribution.

This section is designed to help give you a foundation for leveraging graphical queries in your own work. Hopefully by now you have a rough idea what graphical queries are, how they can be useful, and how to create them with `highlight_key()` and `highlight()`. Understanding the basic idea is one thing, but applying it effectively to new problems is another thing entirely. To help spark your imagination and demonstrate what's possible, Section 16.4 has numerous subsections each with numerous examples of graphical queries in action.

16.2   Highlight versus filter events

Section 16.1 provides an overview of **plotly**'s framework for *highlight* events, but it also supports *filter* events. These events trigger slightly different logic:

- A highlight event dims the opacity of existing marks, then adds an additional graphical layer representing the selection.
- A filter event completely removes existing marks and rescale axes to the remaining data.[6]

Figure 16.7 provides a quick visual depiction in the difference between filter and highlight events. At least currently, filter events must be fired from filter widgets from the **crosstalk** package, and these widgets expect an object of class `SharedData` as input. As it turns out, the `highlight_key()` function, introduced in Section 16.1, creates a `SharedData` instance and is essentially a wrapper for `crosstalk::SharedData$new()`.

```
class(highlight_key(mtcars))
#> [1] "SharedData" "R6"
```

Figure 16.7 demonstrates the main difference in logic between filter and highlight events. Notice how, in the code implementation, the 'querying variable' definition for filter events is part of the filter widget. That is, `city` is defined as the variable of interest in `filter_select()`, not in the creation of `tx`. That is (intentionally) different from the approach

---

[6]When using `ggplotly()`, you need to specify `dynamicTicks = TRUE`.

for highlight events, where the 'querying variable' is a property of the dataset behind the graphical elements.

```r
library(crosstalk)

# generally speaking, use a "unique" key for filter,
# especially when you have multiple filters!
tx <- highlight_key(txhousing)
gg <- ggplot(tx) + geom_line(aes(date, median, group = city))
filter <- bscols(
  filter_select("id", "Select a city", tx, ~city),
  ggplotly(gg, dynamicTicks = TRUE),
  widths = c(12, 12)
)

tx2 <- highlight_key(txhousing, ~city, "Select a city")
gg <- ggplot(tx2) + geom_line(aes(date, median, group = city))
select <- highlight(
  ggplotly(gg, tooltip = "city"),
  selectize = TRUE, persistent = TRUE
)

bscols(filter, select)
```

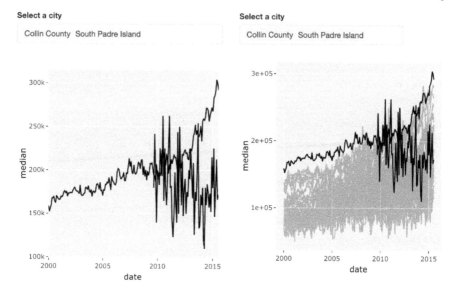

FIGURE 16.7: Comparing filter to highlight events. Filter events completely remove existing marks and rescales axes to the remaining data. For a video demonstration of the interactive, see `https://bit.ly/filter-highlight`. For the interactive, see `https://plotly-r.com/interactives/filter-highlight.html`

When using multiple filter widgets to filter the same dataset, as done in Figure 16.8, you should avoid referencing a non-unique querying variable (i.e., key-column) in the `SharedData` object used to populate the filter widgets. Remember that the default behavior of `highlight_key()` and `SharedData$new()` is to use the row-index (which is unique). This ensures the intersection of multiple filtering widgets queries the correct subset of data.

```
library(crosstalk)
tx <- highlight_key(txhousing)
widgets <- bscols(
  widths = c(12, 12, 12),
  filter_select("city", "Cities", tx, ~city),
  filter_slider("sales", "Sales", tx, ~sales),
  filter_checkbox("year", "Years", tx, ~year, inline = TRUE)
```

```
)
bscols(
  widths = c(4, 8), widgets,
  plot_ly(tx, x = ~date, y = ~median, showlegend = FALSE) %>%
    add_lines(color = ~city, colors = "black")
)
```

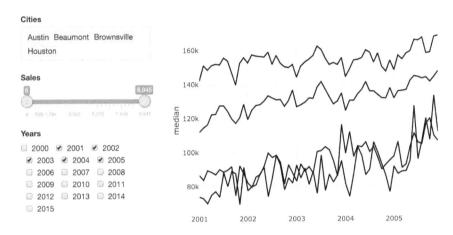

FIGURE 16.8: Filtering on multiple variables. For a video demonstration of the interactive, see https://bit.ly/multiple-filter-widgets. For the interactive, see https://plotly-r.com/interactives/multiple-filter-widgets.html

As Figure 16.9 demonstrates, filter and highlight events can work in conjunction with various **htmlwidgets**. In fact, since the semantics of filter are more well defined than highlight, linking filter events across **htmlwidgets** via **crosstalk** should generally be more well supported.[7]

```
library(leaflet)

eqs <- highlight_key(quakes)
stations <- filter_slider(
```

---

[7]All R packages with **crosstalk** support are currently listed here: https://rstudio.github.io/crosstalk/widgets.html

```
  "station", "Number of Stations",
  eqs, ~stations
)

p <- plot_ly(eqs, x = ~depth, y = ~mag) %>%
  add_markers(alpha = 0.5) %>%
  highlight("plotly_selected")

map <- leaflet(eqs) %>%
  addTiles() %>%
  addCircles()

bscols(
  widths = c(6, 6, 3),
  p, map, stations
)
```

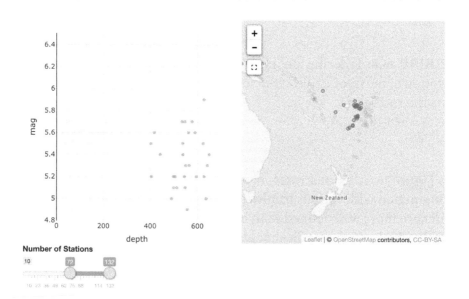

FIGURE 16.9: Linking **plotly** and **leaflet** through both **filter** and **highlight** events. For a video demonstration of the interactive, see https:// bit.ly/plotly-leaflet-filter. For the interactive, see https://plotly-r.com/interactives/plotly-leaflet-filter.html

When combining filter and highlight events, one (current) limitation to be aware of is that the highlighting variable has to be nested inside filter variable(s). For example, in Figure 16.10, we can filter by continent and highlight by country, but there is currently no way to highlight by continent and filter by country.

```r
library(gapminder)
g <- highlight_key(gapminder, ~country)
continent_filter <- filter_select(
  "filter", "Select a country",
  g, ~continent
)

p <- plot_ly(g) %>%
  group_by(country) %>%
  add_lines(x = ~year, y = ~lifeExp, color = ~continent) %>%
  layout(xaxis = list(title = "")) %>%
  highlight(selected = attrs_selected(showlegend = FALSE))

bscols(continent_filter, p, widths = 12)
```

Select a country

Africa Europe

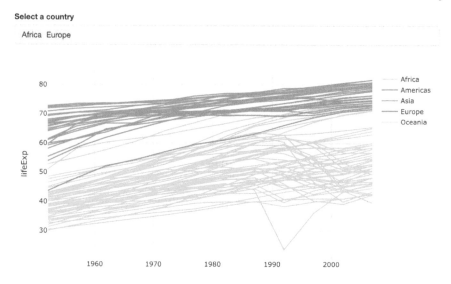

FIGURE 16.10: Combining filtering and highlighting with non-unique querying variables. For a video demonstration of the interactive, see `https://bit.ly/gapminder-filter-highlight`. For the interactive, see `https://plotly-r.com/interactives/gapminder-filter-highlight.html`

16.3   Linking animated views

The graphical querying framework (Section 16.1) works in tandem with key-frame animations Chapter 14. Figure 16.11 extends Figure 14.1 by layering on linear models specific to each frame and specifying `continent` as a key variable. As a result, one may interactively highlight any continent they wish, and track the relationship through the animation. In the animated version of Figure 14.1, the user highlights the Americas, which makes it much easier to see that the relationship between GDP per capita and life expectancy was very strong starting in the 1950s, but progressively weakened throughout the years.

```
g <- highlight_key(gapminder, ~continent)
gg <- ggplot(g, aes(gdpPercap, lifeExp,
  color = continent, frame = year)) +
```

```
geom_point(aes(size = pop, ids = country)) +
geom_smooth(se = FALSE, method = "lm") +
scale_x_log10()
highlight(ggplotly(gg), "plotly_hover")
```

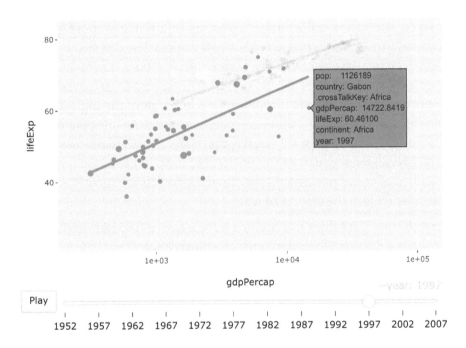

FIGURE 16.11: Highlighting the relationship between GDP per capita and life expectancy in the Americas and tracking that relationship through several decades. For a video demonstration of the interactive, see https://bit.ly/gapminder-highlight-animation. For the interactive, see https://plotly-r.com/interactives/gapminder-highlight-animation.html

In addition to highlighting objects within an animation, objects may also be linked between animations. Figure 16.12 links two animated views: on the left-hand side is population density by country, and on the right-hand side is GDP per capita versus life expectancy. By default, all of the years are shown in black and the current year is shown in red. By pressing play to animate through the years, we can see that all three of these variables have increased (on average) fairly consistently over time. By linking the animated layers, we may condition on an interesting region of this data space to make comparisons in the overall relationship over time.

For example, in Figure 16.12, countries below the 50th percentile in terms of population density are highlighted in blue, then the animation is played again to reveal a fairly interesting difference in these groups. From 1952 to 1977, countries with a low population density seem to enjoy large increases in GDP per capita and moderate increases in life expectancy, then in the early 80s, their GPD seems to decrease while the life expectancy greatly increases. In comparison, the high-density countries seem to enjoy a more consistent and steady increase in both GDP and life expectancy. Of course, there are a handful of exceptions to the overall trend, such as the noticeable drop in life expectancy for a handful of countries during the nineties, which are mostly African countries feeling the effects of war.

The gapminder data does not include a measure of population density, but the gap dataset (included with the **plotlyBook** R package) adds a column containing the population per square kilometer (popDen), which helps implement Figure 16.12. In order to link the animated layers (i.e., red points), we need another version of gap that marks the country variable as the link between the plots (gapKey).

```
data(gap, package = "plotlyBook")

gapKey <- highlight_key(gap, ~country)

p1 <- plot_ly(gap, y = ~country, x = ~popDen, hoverinfo = "x") %>%
  add_markers(alpha = 0.1, color = I("black")) %>%
  add_markers(
    data = gapKey,
    frame = ~year,
    ids = ~country,
    color = I("red")
  ) %>%
```

```
  layout(xaxis = list(type = "log"))


p2 <- plot_ly(gap, x = ~gdpPercap, y = ~lifeExp, size = ~popDen,
                 text = ~country, hoverinfo = "text") %>%
  add_markers(color = I("black"), alpha = 0.1) %>%
  add_markers(
    data = gapKey,
    frame = ~year,
    ids = ~country,
    color = I("red")
  ) %>%
  layout(xaxis = list(type = "log"))


subplot(p1, p2, nrows = 1, widths = c(0.3, 0.7), titleX = TRUE) %>%
  hide_legend() %>%
  animation_opts(1000, redraw = FALSE) %>%
  layout(hovermode = "y", margin = list(l = 100)) %>%
  highlight(
    "plotly_selected",
    color = "blue",
    opacityDim = 1,
    hoverinfo = "none"
  )
```

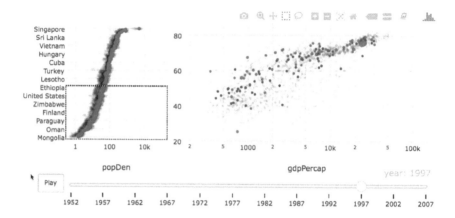

FIGURE 16.12: Comparing the evolution in the relationship between per capita GDP and life expectancy in countries with large populations (red) and small populations (blue). For a video demonstration of the interactive, see https://bit.ly/animation-gapminder. For the interactive, see https://plotly-r.com/interactives/animation-gapminder.html

16.4   Examples

16.4.1   Querying faceted charts

A faceted chart, also known as a trellis or small multiples display, is an effective way to observe how a certain relationship or visual pattern changes with a discrete variable (Becker et al., 1996) (Tufte, 2001). The implementation of a faceted chart partitions a dataset into groups, then produces a graphical panel for each group using a fixed visual encoding (e.g., a scatterplot). When these groups are related in some way, it can be useful to consider linking the panels through graphical queries to reveal greater insight, especially when it comes to making comparisons both within and across multiple groups.

Figure 16.13 is an example of making comparisons both within and across panels via graphical querying in a faceted chart. Each panel represents one year of English Premier League standings across time, and each line represents a team (the querying variable). Since the xaxis represents the number of games within season, and the y-axis tracks cumulative points relative to the league average, lines with a positive slope represent above-average performance and a negative slope represents below-average performance. This design makes it easy to query a good (or bad) team for a particular year (via direct manipulation) to see who the team is as well as how it has compared to the competition in other years. In addition, the dynamic and persistent color brush allows us to query other teams to compare both within and across years. This example is shipped as a demo with the **plotly** package and uses data from the **engsoccerdata** package (Curley, 2016). Thanks to Antony Unwin for providing the initial idea and inspiration for Figure 16.13 (Unwin, 2016).

```
# By entering this demo in your R console it will print out the
# actual source code necessary to recreate the graphic
# Also, `demo(package = "plotly")` will list of all
# demos shipped with plotly
demo("crosstalk-highlight-epl-2", package = "plotly")
```

Brush color Select a team

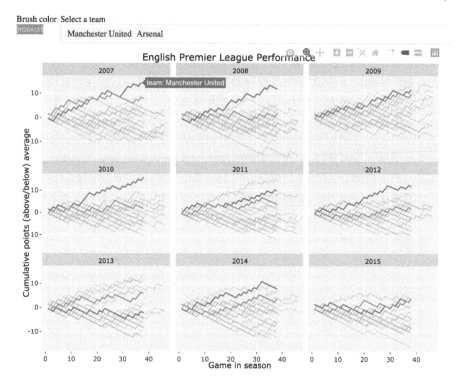

FIGURE 16.13: Graphical querying in a faceted worm chart of English Premier League football teams between 2007 and 2015. The combination of direct and indirect manipulation with the dynamic color brush makes it easy to make comparisons between good and/or bad teams relative to their known rivals. This particular comparison of Man U vs. Arsenal demonstrates that, for the most part, Man U performed better from 2007 to 2015, except in 2013. For a video demonstration of the interactive, see `https://bit.ly/plotly-epl`. For the interactive, see `https://plotly-r.com/interactives/epl.html`

The demo above requires some fairly advanced data pre-processing, so to learn how to implement graphical queries in trellis displays, let's work with more minimal examples. Figure 16.14 gives us yet another look at the `txhousing` dataset. This time we focus on just four cities and give each city its own panel in the trellis display by leveraging `facet_wrap()` from **ggplot2**. Within each panel, we'll wrap the house price time series by year by putting the month on the x-axis and grouping by year. Then, to link these panels, we'll assign year as a querying variable. As a result, not only do we have the ability to analyze annual trends within city, but we can also query specific years to compare unusual or interesting years both within and across cities.

```r
library(dplyr)
cities <- c("Galveston", "Midland", "Odessa", "South Padre Island")
txsmall <- txhousing %>%
  select(city, year, month, median) %>%
  filter(city %in% cities)

txsmall %>%
  highlight_key(~year) %>% {
    ggplot(., aes(month, median, group = year)) + geom_line() +
      facet_wrap(~city, ncol = 2)
  } %>%
  ggplotly(tooltip = "year")
```

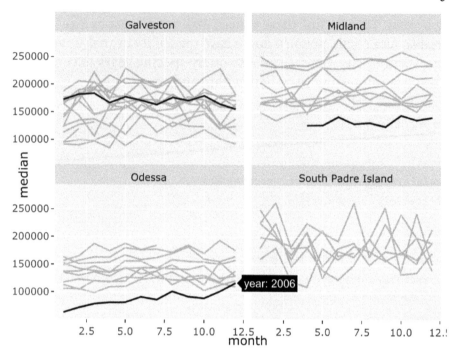

FIGURE 16.14: Monthly median house prices in four Texan cities. Querying by year allows one to compare unusual or interesting years both within and across cities.

Figure 16.15 displays the same information as 16.14 but shows a way to implement a linked trellis display via plot_ly() instead of ggplotly(). This approach leverages dplyr::do() to create **plotly** object for each city/panel, then routes that list of plots into subplot(). One nuance here is that the querying variable has to be defined within the do() statement, but every time highlight_key() is called, it creates a crosstalk::SharedData object belonging to a new unique group, so to link these panels together, the group must be set to a constant value (here we've set group = "txhousing-trellis").

```
txsmall %>%
  group_by(city) %>%
  do(
    p = highlight_key(., ~year, group = "txhousing-trellis") %>%
```

```
    plot_ly(showlegend = FALSE) %>%
    group_by(year) %>%
    add_lines(
      x = ~month, y = ~median, text = ~year,
      hoverinfo = "text"
    ) %>%
    add_annotations(
      text = ~unique(city),
      x = 0.5, y = 1,
      xref = "paper", yref = "paper",
      xanchor = "center", yanchor = "bottom",
      showarrow = FALSE
    )
) %>%
subplot(
  nrows = 2, margin = 0.05,
  shareY = TRUE, shareX = TRUE, titleY = FALSE
)
```

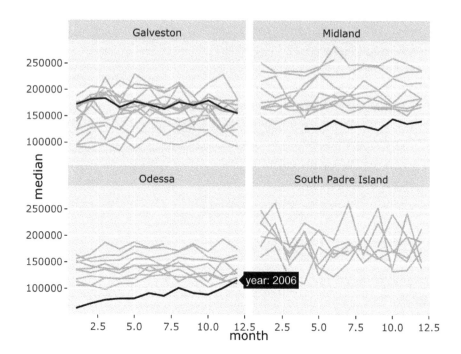

FIGURE 16.15: Using `plot_ly()` instead of `ggplotly()` to implement a linked trellis display.

16.4.2   Statistical queries

16.4.2.1   Statistical queries with `plot_ly()`

Figure 16.6 introduced the concept of leveraging statistical trace types inside the graphical querying framework. This section gives some more examples of leveraging these trace types to dynamically produce statistical summaries of graphical queries. But first, to help understand what makes a trace "statistical", consider the difference between `add_bars()` and `add_histogram()` (described in detail in Chapter 5). The important difference here is that `add_bars()` requires the bar heights to be pre-specified, whereas plotly.js does the relevant computations in `add_histogram()`. More generally, with a statistical trace, you provide a collection of "raw" values and plotly.js performs the statistical summaries necessary to render the graphic. As Figure 16.23 shows, sometimes you'll want to fix certain parameters of the summary (e.g., number of bins in a histogram) to ensure the selection layer is comparable to the original layer.

Figure 16.16 demonstrates routing of a scatterplot brushing event to two different statistical trace types: `add_boxplot()` and `add_histogram()`. Here we've selected all cars with 4 cylinders to show that cylinders appear to have a significant impact on miles per gallon for pickups and sport utility vehicles, but the interactive graphic allows us to query any subset of cars. Often with scatterplot brushing, it's desirable to have the row index inform the SQL query (i.e., have a 1-to-1 mapping between a row of data and the marker encoding that row). This happens to be the default behavior of `highlight_key()`; if no data variable is specified, then it automatically uses the row index as the querying variable.

```
demo("crosstalk-highlight-binned-target-a", package = "plotly")
```

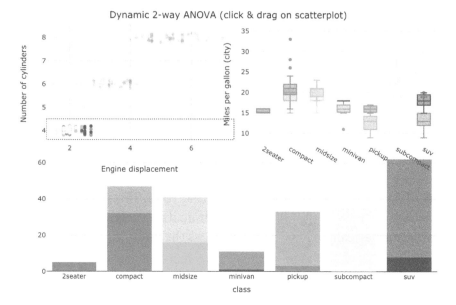

FIGURE 16.16: Linking a (jittered) dotplot of engine displacement by number of cylinders with boxplots of miles per gallon split by class and a bar chart of Dynamic 2-way ANOVA. For a video demonstration of the interactive, see `https://bit.ly/two-way-anova`. For the interactive, see `https://plotly-r.com/interactives/2-way-anova.html`

When using a statistical trace type with graphical queries, it's often desirable to set the querying variable as the row index. That's because, with a statistical trace, numerous data values are attached to each graphical mark; and in that case, it's most intuitive if each value queries just one observation. Figure 16.17 gives a simple example of linking a (dynamic) bar chart with a scatterplot in this way to allow us to query interesting regions of the data space defined by engine displacement (`disp`), miles per gallon on the highway (`hwy`), and the class of car (`class`). Notice how selections can derive from either view, and since we've specified `"plotly_selected"` as the on event, either rectangular or lasso selections can be used to trigger the query.

```
d <- highlight_key(mpg)
base <- plot_ly(d, color = I("black"), showlegend = FALSE)
```

```
subplot(
  add_histogram(base, x = ~class),
  add_markers(base, x = ~displ, y = ~hwy)
) %>%
  # Selections are actually additional traces, and, by default,
  # plotly.js will try to dodge bars placed under the same category
  layout(barmode = "overlay", dragmode = "lasso") %>%
  highlight("plotly_selected")
```

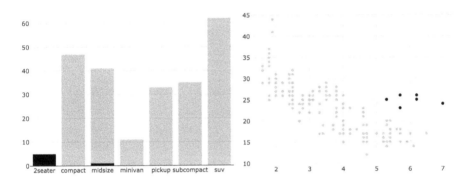

FIGURE 16.17: Linking a bar chart with a scatterplot to query interesting regions of the data space defined by engine displacement (`disp`), miles per gallon highway (`hwy`), and the class of car (`class`). Notice how, by using `add_histogram()`, the number of cars within each class is dynamically computed by plotly.js. For a video demonstration of the interactive, see `https://bit.ly/mpg-linked-bars`. For the interactive, see `https://plotly-r.com/interactives/mpg-linked-bars.html`

Figure 16.18 adds two more statistical trace types to Figure 16.17 to further explore how miles per gallon highway is related to fuel type (`fl`) and front/rear/4 wheel drive (`drv`). In particular, one can effectively condition on these discrete variables to see how the other distributions respond by brushing and dragging over markers. For example, in Figure 16.18, front-wheel drive cars are highlighted in red, then 4-wheel drive cars in blue, and as a result, we can see a large main effect of going from 4 to front-wheel drive. Moreover, among these categories, there are large interactions with regular and diesel fuel types (i.e., given

you have a diesel engine, there is a huge difference between front and
4-wheel drive).

```r
d <- highlight_key(mpg)
base <- plot_ly(d, color = I("black"), showlegend = FALSE)

subplot(
  add_markers(base, x = ~displ, y = ~hwy),
  add_boxplot(base, x = ~fl, y = ~hwy) %>%
    add_markers(x = ~fl, y = ~hwy, alpha = 0.1),
  add_trace(base, x = ~drv, y = ~hwy, type = "violin") %>%
    add_markers(x = ~drv, y = ~hwy, alpha = 0.1),
  shareY = TRUE
) %>%
  subplot(add_histogram(base, x = ~class), nrows = 2) %>%
  # Selections are actually additional traces, and, by default,
  # plotly.js will try to dodge bars placed under the same category
  layout(barmode = "overlay") %>%
  highlight("plotly_selected", dynamic = TRUE)
```

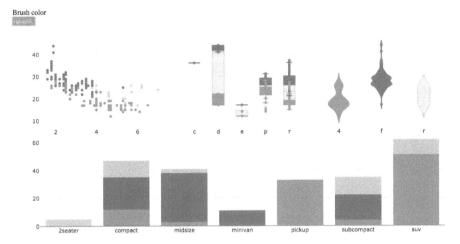

Brush color

FIGURE 16.18: Using statistical queries to perform a 2-way ANOVA on the mpg dataset. Cars with front-wheel drive are highlighted in red, and 4-wheel drive are highlighted in blue. The dynamically rendered boxplots by fuel type indicate significant interaction effects. For a video demonstration of the interactive, see https://bit.ly/mpg-linked. For the interactive, see https://plotly-r.com/interactives/mpg-linked. html

16.4.3   Statistical queries with ggplotly()

Compared to plot_ly(), statistical queries (client-side) with ggplotly() are fundamentally limited. That's because, the statistical R functions that **ggplot2** relies on to generate the graphical layers can't necessarily be recomputed with different input data in your web browser. That being said, this is really only an issue when attempting to *target* a **ggplot2** layer with a non-identity statistic (e.g., geom_smooth(), stat_summary(), etc.). In that case, one should consider linking views server-side, as covered in Chapter 17.

As Figure 16.19 demonstrates, you can still have a **ggplot2** layer with a non-identity statistic serving as the *source* of a selection. In that case, ggplotly() will automatically attach all the input values of the querying variable into the creation of the relevant graphical object (e.g., the fitted line). That is why, in the example below, when a fitted line is hovered upon, all the points belonging to that particular group are highlighted, even when the querying variable is the row index.

```
m <- highlight_key(mpg)
p <- ggplot(m, aes(displ, hwy, colour = class)) +
    geom_point() +
    geom_smooth(se = FALSE, method = "lm")
ggplotly(p) %>% highlight("plotly_hover")
```

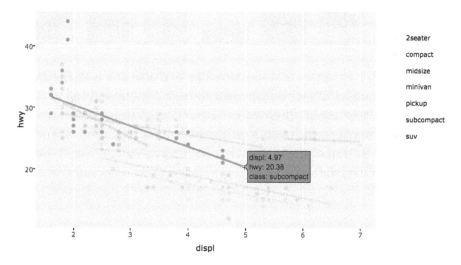

FIGURE 16.19: Engine displacement versus highway miles per gallon by class of car. The linear model for each class, as well as the individual observations, can be selected by hovering over the line of fitted values. An individual observation can also be selected by hovering over the relevant point. For a video demonstration of the interactive, see `https://bit.ly/smooth-highlight`. For the interactive, see `https://plotly-r.com/interactives/smooth-highlight.html`

Figure 16.19 demonstrates highlighting in a single view when the querying variable is the row index, but the linking could also be done by matching the querying variable with the **ggplot2** group of interest, as is done in Figure 16.20. This way, when a user highlights an individual point, the entire group is highlighted (instead of just that one point).

```
m <- highlight_key(mpg, ~class)
p1 <- ggplot(m, aes(displ, fill = class)) + geom_density()
p2 <- ggplot(m, aes(displ, hwy, fill = class)) + geom_point()
subplot(p1, p2) %>% hide_legend() %>% highlight("plotly_hover")
```

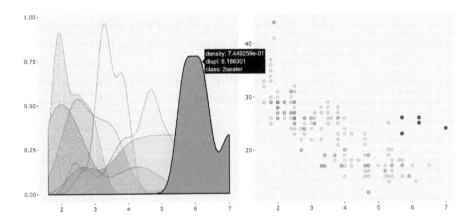

FIGURE 16.20: Clicking on a density estimate to highlight all the raw observations that went into that estimate. For a video demonstration of the interactive, see https://bit.ly/ggplotly-linked-densities. For the interactive, see https://plotly-r.com/interactives/ggplotly-linked-densities.html

In summary, we've learned numerous things about statistical queries:

- A statistical trace (e.g., `add_histogram()`, `add_boxplot()`, etc.) can be used as both the source and target of a graphical query.
- When a statistical trace is the target of a graphical query, it's often desirable to have the row index assigned as the querying variable.
- A **ggplot2** layer can be used as the source of a graphical query, but when it is the target, non-trivial statistical functions cannot be re-computed client-side. In that case, one should consider linking views server-side, as covered in Chapter 17.

16.4.4 Geo-spatial queries

Chapter 4 covers several different approaches[8] for rendering geo-spatial information, and each approach supports graphical querying. One clever approach is to render a 3D globe as a surface, then layer on geo-spatial data on top of that globe with a scatter3d trace. Not only is 3D a nice way to visualize geospatial data that has altitude (in addition to latitude and longitude), but it also grants the ability to interpolate color along a path. Figure 16.21 renders tropical storms paths on a 3D globe and uses color to encode the altitude of the storm at that point. Below the 3D view is a 2D view of altitude versus distance traveled. These views are linked by a graphical query where the querying variable is the storm ID.

```
demo("sf-plotly-3D-globe", package = "plotly")
```

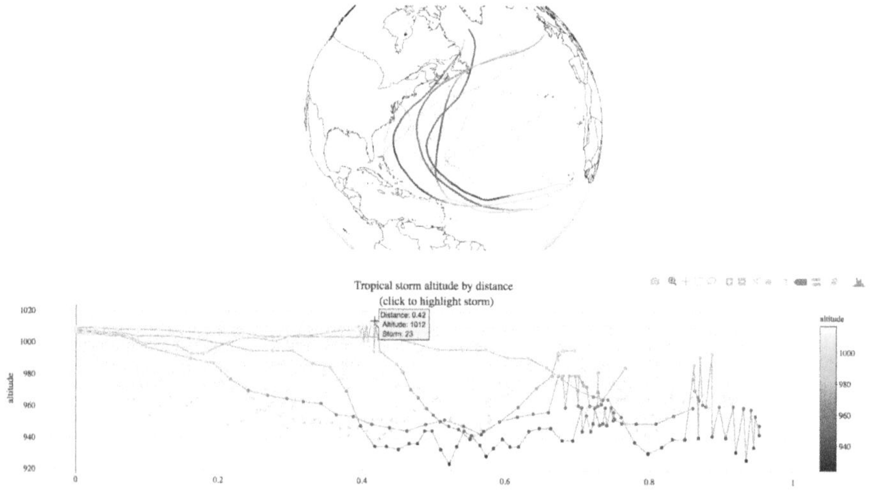

FIGURE 16.21: Linking a 3D globe with tropical storm paths to a 2D view of the storm altitude versus distance traveled. For a video demonstration of the interactive, see https://bit.ly/storms-preview. For the interactive, see https://plotly-r.com/interactives/storms.html

---

[8] Sievert (2018c) outlines the relative strengths and weaknesses of each approach.

A more widely used approach to geo-spatial data visualization is to render lat/lon data on a basemap layer that updates in response to zoom events. The plot_mapbox() function from **plotly** does this via integration with mapbox[9]. Figure 16.22 uses plot_mapbox() highlighting earthquakes west of Fiji to compare the relative frequency of their magnitude and number of reporting stations (to the overall relative frequency).

```r
eqs <- highlight_key(quakes)

# you need a mapbox API key to use plot_mapbox()
# https://www.mapbox.com/signup
map <- plot_mapbox(eqs, x = ~long, y = ~lat) %>%
  add_markers(color = ~depth) %>%
  layout(
    mapbox = list(
      zoom = 2,
      center = list(lon = ~mean(long), lat = ~mean(lat))
    )
  ) %>%
  highlight("plotly_selected")

# shared properties of the two histograms
hist_base <- plot_ly(
    eqs, color = I("black"),
    histnorm = "probability density"
  ) %>%
  layout(barmode = "overlay", showlegend = FALSE) %>%
  highlight(selected = attrs_selected(opacity = 0.5))

histograms <- subplot(
  add_histogram(hist_base, x = ~mag),
  add_histogram(hist_base, x = ~stations),
  nrows = 2, titleX = TRUE
)
```

---

[9] https://www.mapbox.com/

```
crosstalk::bscols(histograms, map)
```

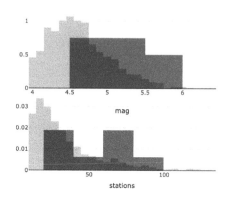 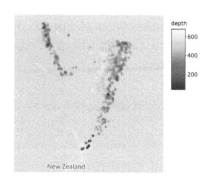

FIGURE 16.22: Querying earthquakes by location and displaying a histogram of their magnitude and number of stations. For a video demonstration of the interactive, see `https://bit.ly/mapbox-quakes`. For the interactive, see `https://plotly-r.com/interactives/mapbox-quakes.html`

Every 2D mapping approach in **plotly** (e.g., `plot_mapbox()`, `plot_ly()`, `geom_sf()`) has a special understanding of the simple features data structure provided by the **sf** package. Sievert (2018b) and Sievert (2018c) go more in depth about simple features support in **plotly** and provide more examples of graphical queries and animation with simple features, but Figure 16.23 demonstrates a clever 'trick' to get bi-directional brushing between polygon centroids and a histogram showing a numerical summary of the polygons. The main idea is to leverage the `st_centroid()` function from **sf** to get the polygons centroids, then link those points to the histogram via `highlight_key()`.

```
library(sf)
nc <- st_read(system.file("shape/nc.shp", package = "sf"))
nc_query <- highlight_key(nc, group = "sf-rocks")
nc_centroid <- highlight_key(st_centroid(nc), group = "sf-rocks")

map <- plot_mapbox(color = I("black"), height = 250) %>%
```

```
  add_sf(data = nc) %>%
  add_sf(data = nc_centroid) %>%
  layout(showlegend = FALSE) %>%
  highlight("plotly_selected", dynamic = TRUE)

hist <- plot_ly(color = I("black"), height = 250) %>%
  add_histogram(
    data = nc_query, x = ~AREA,
    xbins = list(start = 0, end = 0.3, size = 0.01)
  ) %>%
  layout(barmode = "overlay") %>%
  highlight("plotly_selected")

crosstalk::bscols(widths = 12, map, hist)
```

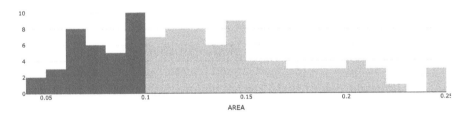

FIGURE 16.23: Graphically querying North Carolina by location and area. For a video demonstration of the interactive, see https://bit.ly/mapbox-bars. For the interactive, see https://plotly-r.com/interactives/mapbox-bars.html

16.4.5   Linking with other htmlwidgets

The **plotly** package is able to share graphical queries with a limited set of other R packages that build upon the **htmlwidgets** standard. At the moment, graphical queries work best with **leaflet** and **DT**. Figure 16.24 links **plotly** with **DT**, and since the dataset linked between the two is an **sf** data frame, each row of the table is linked to a polygon on the map through the row index of the same dataset.

```
demo("sf-dt", package = "plotly")
```

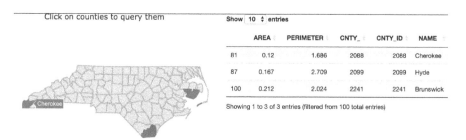

FIGURE 16.24: Linking a `plot_ly()`-based map with a `datatable()` from the **DT** package. For a video demonstration of the interactive, see `https://bit.ly/sf-dt`. For the interactive, see `https://plotly-r.com/interactives/sf-dt.html`

As already shown in Section 16.2, **plotly** can share graphical queries with **leaflet**. Some of the more advanced features (e.g., persistent selection with dynamic color brush) are not yet officially supported, but you can still leverage these experimental features by installing the experimental versions of **leaflet** referenced in the code below. For example, in Figure 16.25, persistent selection with dynamic colors allows one to first highlight earthquakes with a magnitude of 5 or higher in red, then earthquakes with a magnitude of 4.5 or lower, and the corresponding earthquakes are highlighted in the leaflet map. This reveals an interesting relationship in magnitude and geographic location, and **leaflet** provides the ability to zoom and pan on the map to investigate regions that have a high density of quakes.

```
# requires an experimental version of leaflet
# devtools::install_github("rstudio/leaflet#346")
library(leaflet)

qquery <- highlight_key(quakes)

p <- plot_ly(qquery, x = ~depth, y = ~mag) %>%
  add_markers(alpha = 0.5) %>%
  highlight("plotly_selected", dynamic = TRUE)

map <- leaflet(qquery) %>%
  addTiles() %>%
  addCircles()

# persistent selection can be specified via options()
withr::with_options(
  list(persistent = TRUE),
  crosstalk::bscols(widths = c(6, 6), p, map)
)
```

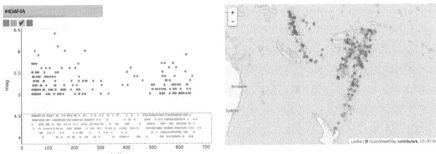

FIGURE 16.25: Linking views between **plotly** and **leaflet** to explore the relation between magnitude and geographic location of earthquakes around Fiji. For a video demonstration of the interactive, see `https://bit.ly/leaflet-persistent`. For the interactive, see `https://plotly-r.com/interactives/leaflet-persistent.html`

Figure 16.26 uses another experimental feature of querying **leaflet** polygons in response to direct manipulation of a **plotly** graph.

```r
# requires an experimental version of leaflet
# devtools::install_github("rstudio/leaflet#391")
library(leaflet)
library(sf)

nc <- system.file("shape/nc.shp", package = "sf") %>%
  st_read() %>%
  st_transform(4326) %>%
  highlight_key()

map <- leaflet(nc) %>%
  addTiles() %>%
  addPolygons(
    opacity = 1,
    color = 'white',
    weight = .25,
    fillOpacity = .5,
    fillColor = 'blue',
    smoothFactor = 0
  )

p <- plot_ly(nc) %>%
  add_markers(x = ~BIR74, y = ~SID79) %>%
  layout(dragmode = "lasso") %>%
  highlight("plotly_selected")

crosstalk::bscols(map, p)
```

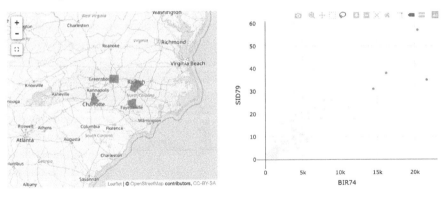

FIGURE 16.26: Querying polygons on a **leaflet** map in response to direct manipulation of a **plotly** graph. For a video demonstration of the interactive, see `https://bit.ly/leaflet-polygons`. For the interactive, see `https://plotly-r.com/interactives/leaflet-polygons.html`

16.4.6 Generalized pairs plots

Section 13.1.2.1 introduced the generalized pairs plot made via `GGally::ggpairs()` which, like `ggplot()`, partially supports graphical queries. The brushing in Figure 16.27 demonstrates how the scatterplots can respond to a graphical queries (allowing us to see how these relationships behave in specific subsections of the data space), but for the same reasons outlined in Section 16.4.3, the statistical summaries (e.g., the density plots and correlations) don't respond to the graphical query.

```
highlight_key(iris) %>%
  GGally::ggpairs(aes(color = Species), columns = 1:4) %>%
  ggplotly() %>%
  highlight("plotly_selected")
```

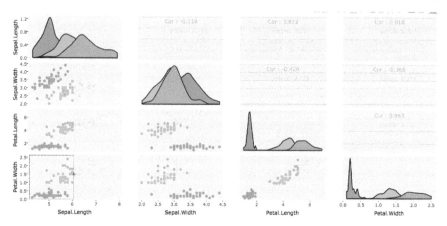

FIGURE 16.27: Brushing a scatterplot matrix via the `ggpairs()` function in the **GGally** package. A video demonstrating the graphical queries can be viewed here `https://bit.ly/linked-ggally`

16.4.7   Querying diagnostic plots

In addition to the `ggpairs()` function for generalized pairs plots, the **GGally** package also has a `ggnostic()` function which generates a matrix of diagnostic plots from a model object using **ggplot2**. Each column of this matrix represents a different explanatory variable and each row represents a different diagnostic measure. Figure 16.28 shows the default display for a linear model, which includes residuals (resid), estimates of residual standard deviation when a particular observation is excluded (sigma), diagonals from the projection matrix (hat), and cooks distance (cooksd).

```
library(dplyr)
library(GGally)

mtcars %>%
  # for better tick labels
  mutate(am = recode(am, `0` = "automatic", `1` = "manual")) %>%
  lm(mpg ~ wt + qsec + am, data = .) %>%
  ggnostic(mapping = aes(color = am)) %>%
  ggplotly()
```

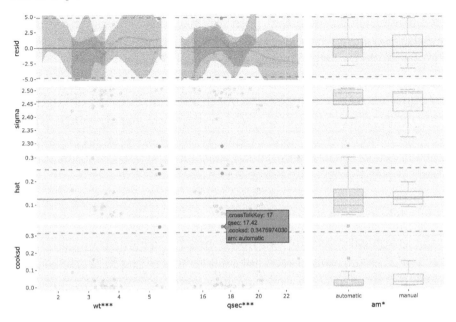

FIGURE 16.28: Graphical queries applied to multiple diagnostic plots of a linear model. The `ggplotly()` function has a special method for `ggnostic()` that adds graphical queries automatically with support for both individual observations (e.g., points) as well as meaningful groups (e.g., automatic vs. manual). For a video demonstration of the interactive, see `https://bit.ly/ggnostic`. For the interactive, see `https://plotly-r.com/interactives/ggnostic.html`

Injecting interactivity into `ggnostic()` via `ggplotly()` enhances the diagnostic plot in at least two ways. Coloring by a factor variable in the model allows us to highlight that region of the design matrix by selecting a relevant statistical summary, which can help avoid overplotting when dealing with numerous factor levels. For example, in Figure 16.28, the user first highlights diagnostics for cars with manual transmission (in blue), then cars with automatic transmission (in red). Perhaps more widely useful is the ability to highlight individual observations since most of these diagnostics are designed to identify highly influential or unusual observations.

In Figure 16.28, there is one observation with a noticeably high value of `cooksd`, which suggests the observation has a large influence on the fitted model. Clicking on that point highlights its corresponding diagnostic measures, plotted against each explanatory variable. Doing so makes it obvious that this observation is influential since it has an unusually high response/residual in a fairly sparse region of the design space (i.e., it has a pretty high value of wt) and removing it would significantly reduce the estimated standard deviation (sigma). By comparison, the other two observations with similar values of wt have a response value very close to the overall mean, so even though their value of hat is high, their value of sigma is low.

16.4.7.1 Subset queries via list-columns

All the graphical querying examples thus far use highlight_key() to attach values from atomic vector of a data frame to graphical marker(s), but what non-atomic vectors (i.e., list-columns)? When it comes to *emitting* events, there is no real difference; **plotly** will "inform the world" of a set of selection values, which is the union of all data values in the graphical query. However, as Figure 16.29 demonstrates, when **plotly** receives a list-column query, it will highlight graphical markers with data value(s) that are a subset of the selected values. For example, when the point [3, 3] is queried, **plotly** will highlight all markers that represent a subset of {A, B, C}, which is why both [1, 1] (representing the set {A}) and (2, 2) (representing the set {A, B}) are highlighted.

```r
d <- tibble::tibble(
  x = 1:4,
  y = 1:4,
  key = lapply(1:4, function(x) LETTERS[seq_len(x)]),
  txt = sapply(key, function(x) {
    sprintf("{%s}", paste(x, collapse = ", "))
  })
)
highlight_key(d, ~key) %>%
  plot_ly(x = ~x, y = ~y, text = ~txt, hoverinfo = "text") %>%
  highlight("plotly_selected", color = "red") %>%
  layout(dragmode = "lasso")
```

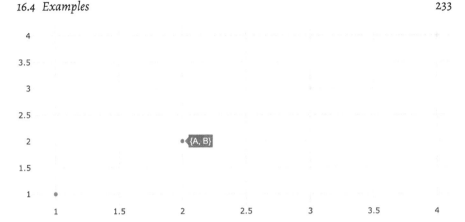

FIGURE 16.29: A simple example of subset queries via a list-column. For a video demonstration of the interactive, see `https://bit.ly/list-column-simple`. For the interactive, see `https://plotly-r.com/interactives/list-column-simple.html`

One compelling use case for subset queries is dendrograms. In fact, **plotly** provides a `plot_dendro()` function for making dendrograms with support for subset queries. Figure 16.30 gives an example of brushing a branch of a dendrogram to query leafs that are similar in some sense. Any dendrogram object can be provided to `plot_dendro()`, but this particular example visualizes the similarity of U.S. states in terms of their arrest statistics via a hierarchical clustering model on the USArrests dataset.

```
hc <- hclust(dist(USArrests), "ave")
dend1 <- as.dendrogram(hc)
plot_dendro(dend1, height = 600) %>%
  hide_legend() %>%
  highlight("plotly_selected", persistent = TRUE, dynamic = TRUE)
```

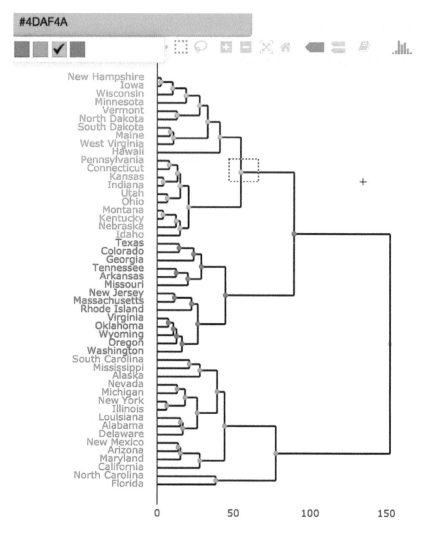

FIGURE 16.30: Leveraging hierarchical selection and persistent brushing to paint branches of a dendrogram. For a video demonstration of the interactive, see https://bit.ly/plotly-dendro. For the interactive, see https://plotly-r.com/interactives/dendro.html

Figure 16.31 links the dendrogram from Figure 16.30 to a map of the U.S. and a grand tour of the arrest statistics to better understand and diagnose a hierarchical clustering methodology. By highlighting branches of the dendrogram, we can effectively choose a partitioning of the states into similar groups, and see how that model choice projects to the data space[10] through a grand tour. The grand tour is a special kind of animation that interpolates between random 2D projections of numeric data allowing the viewer to perceive the shape of a high-dimensional point cloud (Asimov, 1985). Note how the grouping portrayed in Figure 16.31 does a fairly good job of staying separated in the grand tour.

```
demo("animation-tour-USArrests", package = "plotly")
```

---

[10]Typically statistical models are diagnosed by visualizing data in the model space rather than model(s) in the data space. As Wickham et al. (2015) points, the latter approach can be a very effective way to better understand and diagnosis statistical models.

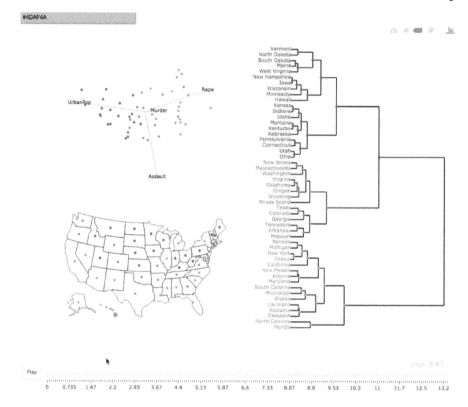

FIGURE 16.31: Linking a dendrogram to a grand tour and map of the USArrests data to visualize a classification in five dimensions. For a video demonstration of the interactive, see https://bit.ly/tour-USArrests. For the interactive, see https://plotly-r.com/interactives/tour-USArrests.html

16.5 Limitations

The graphical querying framework presented here is for posing database queries between multiple graphs via direct manipulation. For serious statistical analysis, one often needs to link other data views (i.e., text-based summaries, tables, etc.) in other arbitrary ways. For these use cases, the R package **shiny** makes it very easy to build on concepts we've already covered to build more powerful client-server applications entirely in R, without having to learn any HTML, CSS, or JavaScript. The next Chapter 17 gives a brief introduction to **shiny**, then dives right into concepts related to linking plotly graphics to other arbitrary views.

17

Server-side linking with shiny

Section 16.1 covers an approach to linking views client-side with graph-ical database queries, but not every linked data view can be reasonably framed as a database query. If you need more control, you have at least two more options: add custom JavaScript (covered in Chapter 18) and/or link views server-side via a web application. Some concepts use-ful for the former approach are covered in Chapter 18, but this chapter is all about the latter approach.

There are several different frameworks for creating web applications via R, but we'll focus our attention on linking **plotly** graphs with **shiny**, an R package for creating reactive web applications entirely in R. **Shiny**'s reactive programming model allows R programmers to build upon their existing R knowledge and create data-driven web appli-cations without any prior web programming experience. **Shiny** itself is largely agnostic to the engine used to render data views (that is, you can incorporate any sort of R output), but **shiny** itself also adds some special support for interacting with static R graphics and images (Chang, 2017).

When linking graphics in a web application, there are tradeoffs to con-sider when using static R plots over web-based graphics. As it turns out, those tradeoffs complement nicely with the relative strengths and weaknesses of linking views with **plotly**, making their combination a powerful toolkit for linking views on the web from R. **Shiny** itself pro-vides a way to access events with static graphics made with any of the following R packages: **graphics**, **ggplot2**, and **lattice**. These packages are very mature, fully featured, well-tested, and support an incredibly wide range of graphics, but since they must be regenerated on the server, they are fundamentally limited from an interactive graphics perspective. Comparatively speaking, **plotly** does not have the same range and history, but it does provide more options and control over interactivity. More specifically, because **plotly** is inherently web-based, it allows for more control over how the graphics update in response to user input (e.g., change the color of a few points instead of redrawing the entire image). This idea is explored in more depth in Section 17.3.1.

This chapter teaches you how to use **plotly** graphs inside **shiny**, how to get those graphics communicating with other types of data views, and how to do it all efficiently. Section 17.1 provides an introduction to **shiny** its reactive programming model, Section 17.2 shows how to leverage **plotly** inputs in **shiny** to coordinate multiple views, Section 17.3.1 shows how to respond to input changes efficiently, and Section 17.4 demonstrates some advanced applications.

17.1 Embedding plotly in shiny

Before linking views with **plotly** inside **shiny**, let's first talk about how to embed **plotly** inside a basic **shiny** app! Through a couple of basic examples, you'll learn the basic components of a **shiny** and get a feel for **shiny**'s reactive programming model, as well as pointers to more learning materials.

17.1.1 Your first shiny app

The most common **plotly+shiny** pattern uses a **shiny** input to control a **plotly** output. Figure 17.1 gives a simple example of using **shiny**'s `selectizeInput()` function to create a dropdown that controls a **plotly** graph. This example, as well as every other **shiny** app, has two main parts:

1. The *user interface*, `ui`, defines how inputs and output widgets are displayed on the page. The `fluidPage()` function offers a nice and quick way to get a grid-based responsive layout[1], but

---

[1]Read more about **shiny**'s responsive layout here `https://shiny.rstudio.com/articles/layout-guide.html`

it's also worth noting the UI is completely customizable[2], and packages such as **shinydashboard** make it easy to leverage more sophisticated layout frameworks (Chang and Borges Ribeiro, 2018).

2. The *server* function, server, defines a mapping from input values to output widgets. More specifically, the **shiny** server is an R function() between input values on the client and outputs generated on the web server.

Every input widget, including the selectizeInput() in Figure 17.1, is tied to an input value that can be accessed on the server inside a reactive expression. **Shiny**'s reactive expressions build a dependency graph between outputs (aka, reactive endpoints) and inputs (aka, reactive sources). The true power of reactive expressions lies in their ability to chain together and cache computations, but let's first focus on generating outputs. In order to generate an output, you have to choose a suitable function for rendering the result of a reactive expression.

Figure 17.1 uses the renderPlotly() function to render a reactive expression that generates a **plotly** graph. This expression depends on the input value input$cities (i.e., the input value tied to the input widget with an inputId of "cities") and stores the output as output$p. This instructs **shiny** to insert the reactive graph into the plotlyOutput(outputId = "p") container defined in the user interface.

```
library(shiny)
library(plotly)

ui <- fluidPage(
  selectizeInput(
    inputId = "cities",
    label = "Select a city",
    choices = unique(txhousing$city),
    selected = "Abilene",
```

---

[2]Read more about using custom HTML templates here https://shiny.rstudio.com/articles/html-ui.html

```
    multiple = TRUE
  ),
  plotlyOutput(outputId = "p")
)


server <- function(input, output, ...) {
  output$p <- renderPlotly({
    plot_ly(txhousing, x = ~date, y = ~median) %>%
      filter(city %in% input$cities) %>%
      group_by(city) %>%
      add_lines()
  })
}


shinyApp(ui, server)
```

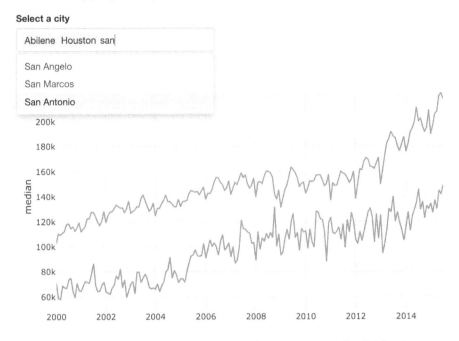

FIGURE 17.1: Using a **shiny** input widget to control which time series are shown on a **plotly** graph. For a video demonstration of the interactive, see `https://bit.ly/shiny-intro`. For the interactive, see `https://plotly-r.com/interactives/shiny-intro.html`

If, instead of a **plotly** graph, a reactive expression generates a static R graphic, simply use `renderPlot()` (instead of `renderPlotly()`) to render it and `plotOutput()` (instead of `plotlyOutput()`) to position it. Other **shiny** output widgets also use this naming convention: `renderDataTable()`/`datatableOutput()`, `renderPrint()`/`verbatimTextOutput()`, `renderText()`/`textOutput()`, `renderImage()`/`imageOutput()`, etc. Packages that are built on the **htmlwidgets** standard (e.g., **plotly** and **leaflet**) are, in some sense, also **shiny** output widgets that are encouraged to follow this same naming convention (e.g., `renderPlotly()`/`plotlyOutput()` and `renderLeaflet()`/`leafletOutput()`).

Shiny also comes pre-packaged with a handful of other useful input widgets. Although many **shiny** apps use them straight "out-of-the-box", input widgets can easily be stylized with CSS and/or SASS, and even custom input widgets can be integrated (Mastny, 2018; RStudio, 2014a).

- `selectInput()`/`selectizeInput()` for dropdown menus.
- `numericInput()` for a single number.
- `sliderInput()` for a numeric range.
- `textInput()` for a character string.
- `dateInput()` for a single date.
- `dateRangeInput()` for a range of dates.
- `fileInput()` for uploading files.
- `checkboxInput()`/`checkboxGroupInput()`/`radioButtons()` for choosing a list of options.

Going forward, our focus is to link multiple graphs in **shiny** through direct manipulation, so we focus less on using these input widgets, and more on using **plotly** and static R graphics as inputs to other output widgets. Section 17.2 provides an introduction to this idea, but before we learn how to access these input events, you may want to know a bit more about rendering **plotly** inside **shiny**.

17.1.2 Hiding and redrawing on resize

The `renderPlotly()` function renders anything that the `plotly_build()` function understands, including `plot_ly()`, `ggplotly()`, and **ggplot2** objects.[3] It also renders NULL as an empty HTML div, which is handy for certain cases where it doesn't make sense to render a graph. Figure 17.2 leverages these features to render an empty div while `selectizeInput()`'s placeholder is shown, but then render a **plotly** graph via `ggplotly()` once cities have been selected. Figure 17.2 also shows how to make the **plotly** output depend on the size of the container that holds the **plotly** graph. By default, when a browser is resized, the graph size is changed purely client-side, but this reactive expression will re-execute when the browser window is resized. Due to technical

---

[3]The `plotly_build()` function is an S3 generic, so you can list all relevant methods with `methods(plotly_build)`, and write your own method to translate a custom object to **plotly**.

reasons, this can improve `ggplotly()` resizing behavior[4], but should be used with caution when handling large data and long render times.

```r
library(shiny)

cities <- unique(txhousing$city)

ui <- fluidPage(
  selectizeInput(
    inputId = "cities",
    label = NULL,
    # placeholder is enabled when 1st choice is an empty string
    choices = c("Please choose a city" = "", cities),
    multiple = TRUE
  ),
  plotlyOutput(outputId = "p")
)

server <- function(input, output, session, ...) {
  output$p <- renderPlotly({
    req(input$cities)
    if (identical(input$cities, "")) return(NULL)
    p <- ggplot(data = filter(txhousing, city %in% input$cities)) +
      geom_line(aes(date, median, group = city))
    height <- session$clientData$output_p_height
    width <- session$clientData$output_p_width
    ggplotly(p, height = height, width = width)
  })
}


shinyApp(ui, server)
```

---

[4]In order to convert **grid** grobs that are relatively sized, the `ggplotly()` function uses the size of the current graphics device at print-time, meaning that resizing the browser window without a hook back to R can create wonky sizes.

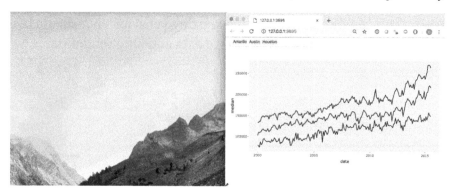

FIGURE 17.2: Rendering a **plotly** graph in **shiny** if and only if the selectizeInput()'s dropdown is non-empty. When the graph is present, and the window is resized, then the reactive expression is re-evaluated. For a video demonstration of the interactive, see https://bit.ly/shiny-ggplotly. For the interactive, see https://plotly-r.com/interactives/shiny-ggplotly.html

When a reactive expression inside renderPlotly() re-executes, it triggers a full redraw of the **plotly** graph on the client. Generally speaking, this makes your **shiny** app logic easy to reason about, but it's not always performant enough. For example, say you have a scatterplot with tens of thousands of points, and you just want to add a fitted line to those points (in response to input event)? Instead of redrawing the whole plot from scratch, it can be way more performant to partially update specific components of the visual. Section 17.3.1 covers this idea through a handful of examples.

17.2   Leveraging plotly input events

Section 17.1 covered how to render **shiny** output widgets (e.g., plotlyOutput()) that depend on an input widget, but what about having an output act like an input to another output? For example, say we'd like to dynamically generate a bar chart (i.e., an output) based on a point clicked on a scatterplot (i.e., an input event tied to an output widget). In addition to **shiny**'s static graph and image rendering functions (e.g., `plotOutput()`/`imageOutput()`), there are a handful of other R packages that expose user interaction with "output" widget(s) as input value(s). Cheng (2018c) and Xie (2018) describe the interface for the **leaflet** and **DT** packages. This section outlines the interface for `plotlyOutput()`. This sort of functionality plays a vital role in linking of views through direct manipulation, similar to what we've already seen in Section 16.1, but having access to **plotly** events on a **shiny** server allows for much more flexibility than linking views purely client-side.

The `event_data()` function is the most straightforward way to access **plotly** input events in **shiny**. Although `event_data()` is function, it references and returns a **shiny** input value, so `event_data()` needs to be used inside a reactive context. Most of these available events are data-specific traces (e.g., `"plotly_hover"`, `"plotly_click"`, `"plotly_selected"`, etc.), but there are also some that are layout-specific (e.g., `"plotly_relayout"`). Most plotly.js events[5] are accessible through this interface; for a complete list, see the `help(event_data)` documentation page.

Numerous figures in the following sections show how to access common **plotly** events in **shiny** and do something with the result. When using these events to inform another view of the data, it's often necessary to know what portion of data was queried in the event (i.e., the x/y positions alone may not be enough to uniquely identify the information of interest). For this reason, it's often a good idea to supply a `key` (and/or `customdata`) attribute, so that you can map the event data back to the original data. The `key` attribute is only supported in **shiny**, but `customdata` is officially supported by plotly.js, and thus can also be used to attach meta-information to event; see Chapter 18 for more details.

17.2.1   Dragging events

There are currently four different modes for mouse click+drag behavior (i.e., `dragmode`) in plotly.js: zoom, pan, rectangular selection, and lasso selection. This mode may be changed interactively via the mode-bar that appears above a **plotly** graph, but the default mode can also be set from the command-line. The default `dragmode` in Figure 17.3

---

[5]These events are documented at https://plot.ly/javascript/plotlyjs-events/

is set to `'select'`, so that dragging draws a rectangular box which highlights markers. When in this mode, or in the lasso selection mode, information about the drag event can be accessed in four different ways: `"plotly_selecting"`, `"plotly_selected"`, `"plotly_brushing"`, and `"plotly_brushed"`. Both the `"plotly_selecting"` and `"plotly_selected"` events emit information about trace(s) appearing within the interior of the brush; the only difference is that `"plotly_selecting"` fires repeatedly *during* drag events, whereas `"plotly_selected"` fires *after* drag events (i.e., after the mouse has been released). The semantics behind `"plotly_brushing"` and `"plotly_brushed"` are similar, but these emit the x/y limits of the selection brush. As for the other two dragging modes (zoom and pan), since they modify the range of the x/y axes, information about these events can be accessed through `"plotly_relayout"`. Sections 17.3.1 and 17.4 both have advanced applications of these dragging events.

```
plotly_example("shiny", "event_data")
```

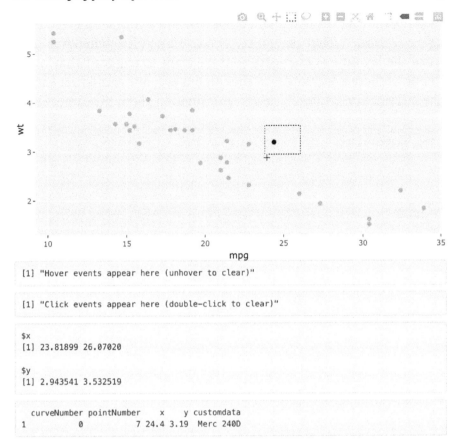

```
[1] "Hover events appear here (unhover to clear)"

[1] "Click events appear here (double-click to clear)"

$x
[1] 23.81899 26.07020

$y
[1] 2.943541 3.532519

  curveNumber pointNumber   x    y customdata
1           0             7 24.4 3.19 Merc 240D
```

FIGURE 17.3: Accessing event data from click and drag events. For a video demonstration of the interactive, see https://bit.ly/plotlyEvents. For the interactive, see https://plotly-r.com/interactives/plotlyEvents.html

17.2.2   3D events

Drag selection events (i.e., "plotly_selecting") are currently only available for 2D charts, but other common events are generally supported for any type of graph, including 3D charts. Figure 17.4 accesses various events in 3D including: "plotly_hover", "plotly_click", "plotly_legendclick", "plotly_legenddoubleclick", and "plotly_relayout". The data emitted via "plotly_hover" and "plotly_click" is structured similarly to data emitted from "plotly_selecting"/"plotly_selected". Figure 17.4 also demonstrates

how one can react to particular components of a conflated event like
`"plotly_relayout"`. That is, `"plotly_relayout"` will fire whenever *any*
part of the layout has changed, so if we want to trigger behavior if
and only if there are changes to the camera eye, one could first check
if the information emitted contains information about the camera eye.

```
plotly_example("shiny", "event_data_3D")
```

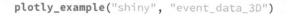

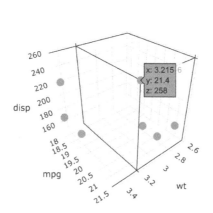

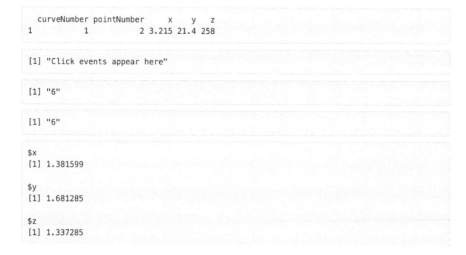

```
    curveNumber pointNumber    x     y   z
1             1           1    2 3.215 21.4 258
```

```
[1] "Click events appear here"
```

```
[1] "6"
```

```
[1] "6"
```

```
$x
[1] 1.381599
```

```
$y
[1] 1.681285
```

```
$z
[1] 1.337285
```

FIGURE 17.4: Accessing 3D events. For a video demonstration of the
interactive, see `https://bit.ly/plotly3Devents`. For the interactive, see
`https://plotly-r.com/interactives/3Devents.html`

17.2.3   Edit events

A little known fact about **plotly** is that you can directly manipulate annotations, title, shapes (e.g., circle, lines, rectangles), legends, and more by simply adding `config(p, editable = TRUE)` to a plot `p`. Moreover, since these are all layout components, we can access and respond to these 'edit events' by listening to the `"plotly_relayout"` events. Figure 17.5 demonstrates how display access information about changes in annotation positioning and content.

```r
library(shiny)

ui <- fluidPage(
  plotlyOutput("p"),
  verbatimTextOutput("info")
)

server <- function(input, output, session) {

  output$p <- renderPlotly({
    plot_ly() %>%
      layout(
        annotations = list(
          list(
            text = emo::ji("fire"),
            x = 0.5,
            y = 0.5,
            xref = "paper",
            yref = "paper",
            showarrow = FALSE
          ),
          list(
            text = "fire",
            x = 0.5,
            y = 0.5,
            xref = "paper",
            yref = "paper"
```

```
          )
        )) %>%
      config(editable = TRUE)
  })

  output$info <- renderPrint({
    event_data("plotly_relayout")
  })

}

shinyApp(ui, server)
```

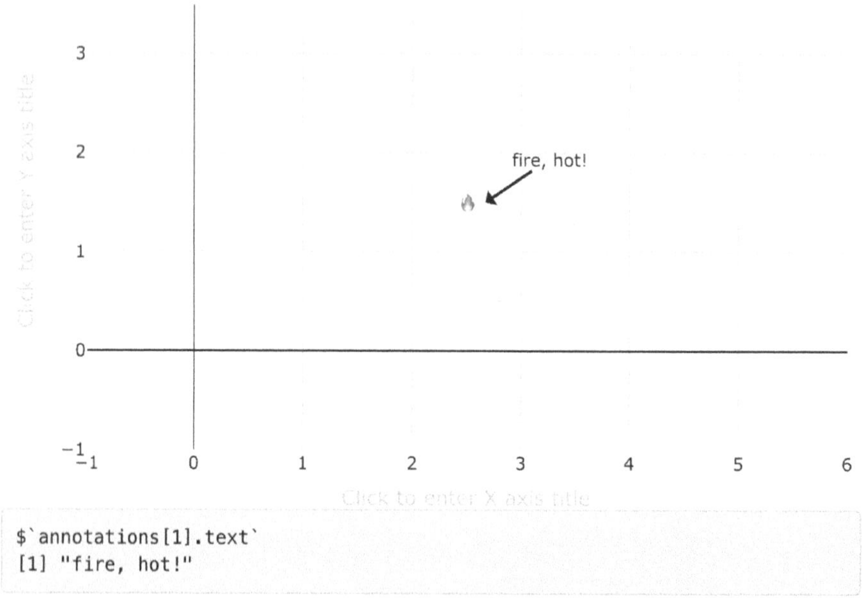

```
$`annotations[1].text`
[1] "fire, hot!"
```

FIGURE 17.5: Accessing information about direct manipulation of annotations. For a video demonstration of the interactive, see `https://bit.ly/shiny-edit-annotations`. For the interactive, see `https://plotly-r.com/interactives/shiny-edit-annotations.html`

Figure 17.6 demonstrates directly manipulating a circle shape and accessing the new positions of the circle. In contrast to Figure 17.5, which

made everything (e.g., the plot and axis titles) editable via `config(p,
editable = TRUE)`, note how Figure 17.6 makes use of the `edits` argument
to make only the shapes editable.

```r
library(shiny)

ui <- fluidPage(
  plotlyOutput("p"),
  verbatimTextOutput("event")
)

server <- function(input, output, session) {

  output$p <- renderPlotly({
    plot_ly() %>%
      layout(
        xaxis = list(range = c(-10, 10)),
        yaxis = list(range = c(-10, 10)),
        shapes = list(
          type = "circle",
          fillcolor = "gray",
          line = list(color = "gray"),
          x0 = -10, x1 = 10,
          y0 = -10, y1 = 10,
          xsizemode = "pixel",
          ysizemode = "pixel",
          xanchor = 0, yanchor = 0
        )
      ) %>%
      config(edits = list(shapePosition = TRUE))
  })

  output$event <- renderPrint({
    event_data("plotly_relayout")
  })
```

```
}

shinyApp(ui, server)
```

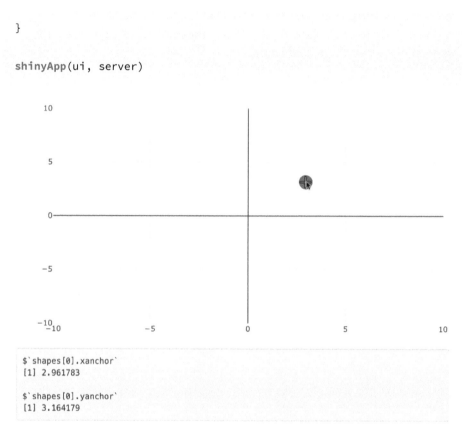

```
$`shapes[0].xanchor`
[1] 2.961783

$`shapes[0].yanchor`
[1] 3.164179
```

FIGURE 17.6: Accessing information about direct manipulation of circle shapes. For a video demonstration of the interactive, see `https://bit.ly/shiny-drag-circle`. For the interactive, see `https://plotly-r.com/interactives/shiny-drag-circle.html`

Figure 17.7 demonstrates a linear model that reacts to edited circle shape positions using the `"plotly_relayout"` event in **shiny**. This interactive tool is an effective way to visualize the impact of high leverage points on a linear model fit. The main idea is to have the model fit (as well as its summary and predicted values) depend on the current state of x and y values, which here is stored and updated via `reactiveValues()`. Section 17.2.8 has more examples of using reactive values to maintain state within a **shiny** application.

```
plotly_example("shiny", "drag_markers")
```

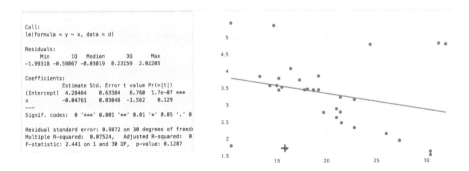

FIGURE 17.7: Editing circle shape positions to dynamically alter a linear model summary and fitted line. This is useful mainly as a teaching device to visually demonstrate the effect of high leverage points on a simple linear model. For a video demonstration of the interactive, see https://bit.ly/interactive-lm. For the interactive, see https://plotly-r.com/interactives/interactive-lm.html

Figure 17.8 uses an editable vertical line and the plotly_relayout event data to 'snap' the line to the closest point in a sequence of x values. It also places a marker on the intersection between the vertical line shape and the line chart of y values. Notice how, by accessing event_data() in this way (i.e., the source and target view of the event is the same), the chart is actually fully redrawn every time the line shape moves. If performance were an issue (i.e., we were dealing with lots of lines), this type of interaction likely won't be very responsive. In that case, you can use event_data() to trigger side effects (i.e., partially modify the plot) which is covered in Section 17.3.1.

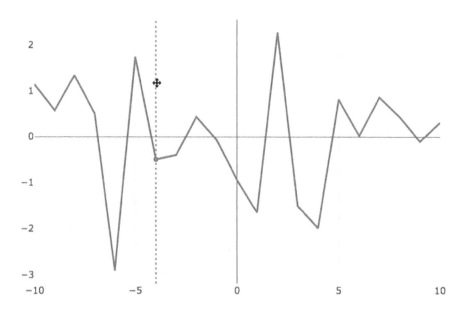

FIGURE 17.8: Dragging a vertical line shape and 'snapping' the line to match the closest provided x value. For a video demonstration of the interactive, see https://bit.ly/shiny-drag-line. For the interactive, see https://plotly-r.com/interactives/shiny-drag-line.html

17.2.4   Relayout vs. restyle events

Remember every **graph** has two critical components: data (i.e., traces) and layout. Similar to how "plotly_relayout" reports partial modifications to the layout, the "plotly_restyle" event reports partial modification to traces. Compared to "plotly_relayout", there aren't very many native direct manipulation events that would trigger a "plotly_restyle" event. For example, zoom/pan events, camera changes, editing annotations/shapes/etc. all trigger a "plotly_relayout" event, but not many traces allow you to directly manipulate their properties. One notable exception is the "parcoords" trace type which has native support for brushing lines along an axis dimension(s). As Figure 17.9 demonstrates,

these brush events emit a `"plotly_restyle"` event with the range(s) of
the highlighted dimension.

```r
library(shiny)

ui <- fluidPage(
  plotlyOutput("parcoords"),
  verbatimTextOutput("info")
)

server <- function(input, output, session) {

  d <- dplyr::select_if(iris, is.numeric)

  output$parcoords <- renderPlotly({

    dims <- Map(function(x, y) {
      list(
        values = x,
        range = range(x, na.rm = TRUE),
        label = y
      )
    }, d, names(d), USE.NAMES = FALSE)

    plot_ly() %>%
      add_trace(
        type = "parcoords",
        dimensions = dims
      ) %>%
      event_register("plotly_restyle")
  })

  output$info <- renderPrint({
    d <- event_data("plotly_restyle")
    if (is.null(d)) "Brush along a dimension" else d
  })
```

```
}
```

```
shinyApp(ui, server)
```

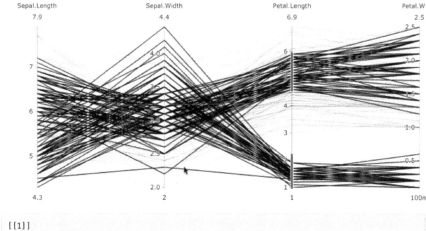

```
[[1]]
[[1]]$`dimensions[2].constraintrange`
, , 1

         [,1]      [,2]
[1,] 0.9539062 4.535175

, , 2

        [,1]  [,2]
[1,] 2.217883 5.963
```

FIGURE 17.9: Using the `"plotly_restyle"` event to access brushed dimensions of a parallel coordinates plot. For a video demonstration of the interactive, see https://bit.ly/shiny-parcoords. For the interactive, see https://plotly-r.com/interactives/shiny-parcoords.html

As Figure 17.10 shows, it's possible to use this information to infer which data points are highlighted. The logic to do so is fairly sophisticated and requires accumulation of the event data, as discussed in Section 17.2.8.

```
plotly_example("shiny", "event_data_parcoords")
```

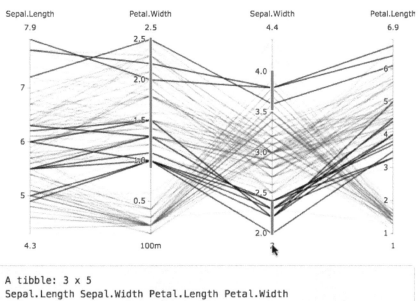

```
# A tibble: 3 x 5
  Sepal.Length Sepal.Width Petal.Length Petal.Width
         <dbl>       <dbl>        <dbl>       <dbl>
1          7.2         3.6          6.1         2.5
2          7.7         3.8          6.7         2.2
3          7.9         3.8          6.4         2
# ... with 1 more variable: Species <fct>
```

FIGURE 17.10: Displaying the highlighted observations of a parcoords trace. For a video demonstration of the interactive, see `https://bit. ly/shiny-parcoords-data`. For the interactive, see `https://plotly-r.com/ interactives/shiny-parcoords-data.html`

17.2.5 Scoping events

This section leverages the interface for accessing **plotly** input events introduced in Section 17.2 to inform other data views about those events. When managing multiple views that communicate with one another, you'll need to be aware of which views are a *source* of interaction and which are a *target* (a view can be both, at once!). The `event_data()` function provides a `source` argument to help refine which view(s) serve as the source of an event. The `source` argument takes a string ID, and when that ID matches the `source` of a `plot_ly()`/`ggplotly()` graph, then the `event_data()` is "scoped" to that view. To get a better idea of how this works, consider Figure 17.11.

Figure 17.11 allows one to click on a cell of correlation heatmap to generate a scatterplot of the two corresponding variables, allowing for a closer look at their relationship. In the case of a heatmap, the event data tied to a `plotly_click` event contains the relevant x and y categories (e.g., the names of the data variables of interest) and the z value (e.g., the pearson correlation between those variables). In order to obtain click data from the heatmap, and only the heatmap, it's important that the `source` argument of the `event_data()` function matches the `source` argument of `plot_ly()`. Otherwise, if the `source` argument was not specified `event_data("plotly_click")` would also fire if and when the user clicked on the scatterplot, likely causing an error.

```r
library(shiny)

# cache computation of the correlation matrix
correlation <- round(cor(mtcars), 3)

ui <- fluidPage(
  plotlyOutput("heat"),
  plotlyOutput("scatterplot")
)

server <- function(input, output, session) {

  output$heat <- renderPlotly({
    plot_ly(source = "heat_plot") %>%
      add_heatmap(
        x = names(mtcars),
        y = names(mtcars),
        z = correlation
      )
  })
```

```r
output$scatterplot <- renderPlotly({
  # if there is no click data, render nothing!
  clickData <- event_data("plotly_click", source = "heat_plot")
  if (is.null(clickData)) return(NULL)

  # Obtain the clicked x/y variables and fit linear model
  vars <- c(clickData[["x"]], clickData[["y"]])
  d <- setNames(mtcars[vars], c("x", "y"))
  yhat <- fitted(lm(y ~ x, data = d))

  # scatterplot with fitted line
  plot_ly(d, x = ~x) %>%
    add_markers(y = ~y) %>%
    add_lines(y = ~yhat) %>%
    layout(
      xaxis = list(title = clickData[["x"]]),
      yaxis = list(title = clickData[["y"]]),
      showlegend = FALSE
    )
})

}

shinyApp(ui, server)
```

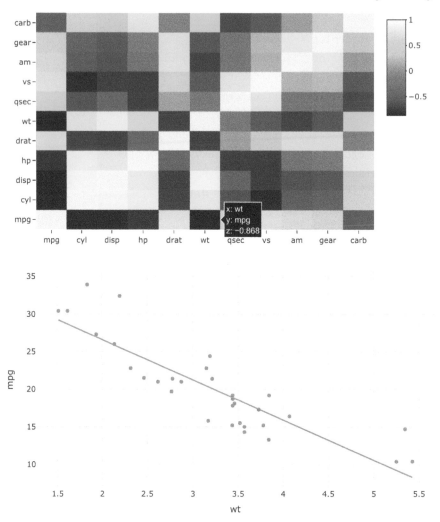

FIGURE 17.11: Linking each cell of a correlation heatmap to the corresponding scatterplot. For a video demonstration of the interactive, see `https://bit.ly/shiny-corrplot`. For the interactive, see `https://plotly-r.com/interactives/shiny-corrplot.html`

17.2.6   Event priority

By default, `event_data()` only invalidates a reactive expression when the value of its corresponding **shiny** input changes. Sometimes, you might want a particular event, say `"plotly_click"`, to *always* invalidate a reactive expression. Figure 17.12 shows the difference between this

default behavior versus setting `priority = 'event'`. By default, repeatedly clicking the same marker won't update the clock, but when setting
the `priority` argument to event, repeatedly clicking the same marker
will update the clock (i.e., it will invalidate the reactive expression).

```r
library(shiny)

ui <- fluidPage(
  plotlyOutput("p"),
  textOutput("time1"),
  textOutput("time2")
)

server <- function(input, output, session) {

  output$p <- renderPlotly({
    plot_ly(x = 1:2, y = 1:2, size = I(c(100, 150)))  %>%
      add_markers()
  })

  output$time1 <- renderText({
    event_data("plotly_click")
    paste("Input priority: ", Sys.time())
  })

  output$time2 <- renderText({
    event_data("plotly_click", priority = "event")
    paste("Event priority: ", Sys.time())
  })

}

shinyApp(ui, server)
```

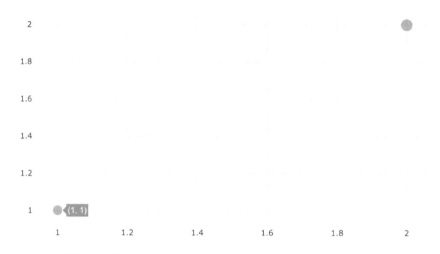

Input priority: 2019-03-31 15:56:37
Event priority: 2019-03-31 15:56:42

FIGURE 17.12: A demo of input priority versus event priority. Clicking on the same marker repeatedly, by default, won't invalidate a reactive expression that depends on 'plotly_click', but it will invalidate when given event priority. For a video demonstration of the interactive, see `https://bit.ly/event-priority`. For the interactive, see `https://plotly-r.com/interactives/event-priority.html`

There are numerous events accessible through `event_data()` that don't contain any information (e.g., `"plotly_doubleclick"`, `"plotly_deselect"`, `"plotly_afterplot"`, etc.). These events are automatically given an event priority since their corresponding **shiny** input value never changes. One common use case for events like `"plotly_doubleclick"` (fired when double-clicking in a zoom or pan dragmode) and `"plotly_deselect"` (fired when double-clicking in a selection mode) is to clear or reset accumulating event data.

17.2.7 Handling discrete axes

For events that are trace-specific (e.g., `"plotly_click"`, `"plotly_hover"`, `"plotly_selecting"`, etc.), the positional data (e.g., x/y/z) is always numeric, so if you have a plot with discrete axes, you might want to know how to map that numeric value back to the relevant input data category.

In some cases, you can avoid the problem by assigning the discrete variable of interest to the `key`/`customdata` attribute, but you might also want to reserve that attribute to encode other information, like a `fill` aesthetic. Figure 17.13 shows how to map the numerical x value emitted in a click event back to the discrete variable that it corresponds to (`mpg$class`) and leverages `customdata` to encode the `fill` mapping allowing us to display the data records a clicked bar corresponds to. In both `ggplotly()` and `plot_ly()`, categories associated with a character vector are always alphabetized, so if you `sort()` the `unique()` character values, then the vector indices will match the x event data values. On the other hand, if x were a factor variable, the x event data would match the ordering of the `levels()` attribute.

```
library(shiny)
library(dplyr)

ui <- fluidPage(
  plotlyOutput("bars"),
  verbatimTextOutput("click")
)

classes <- sort(unique(mpg$class))

server <- function(input, output, session) {

  output$bars <- renderPlotly({
    ggplot(mpg, aes(class, fill = drv, customdata = drv)) +
      geom_bar()
  })

  output$click <- renderPrint({
    d <- event_data("plotly_click")
    if (is.null(d)) return("Click a bar")
    mpg %>%
      filter(drv %in% d$customdata) %>%
      filter(class %in% classes[d$x])
```

```
  })


}


shinyApp(ui, server)
```

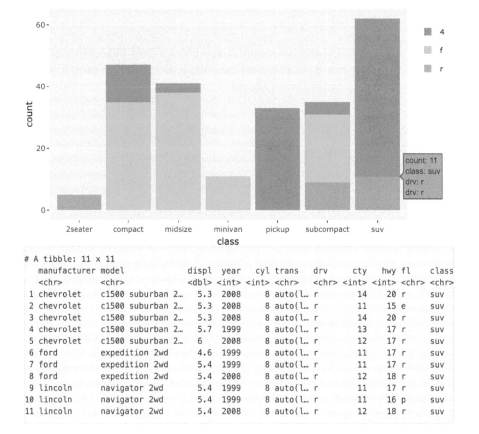

FIGURE 17.13: Retrieving the data observations that correspond to a particular bar in a stacked bar chart. For a video demonstration of the interactive, see https://bit.ly/discrete-event-data. For the interactive, see https://plotly-r.com/interactives/discrete-event-data.html

17.2.8   Accumulating and managing event data

Currently all the events accessible through `event_data()` are *transient*. This means that, given an event like `"plotly_click"`, the value of `event_data()` will only reflect the most recent click information. However, in order to implement complex linked graphics with *persistent* qualities, like Figure 16.3 or 17.28, you'll need some way to accumulate and manage event data. The general mechanism that **shiny** provides to achieve this kind of task is `reactiveVal()` (or, the plural version, `reactiveValues()`), which essentially provides a way to create and manage input values entirely server-side.

Figure 17.14 demonstrates a shiny app that accumulates hover information and paints the hovered points in red. Every time a hover event is triggered, the corresponding car name is added to the set of selected cars, and every time the plot is double-clicked that set is cleared. This general pattern of initializing a reactive value (i.e., `cars <- reactiveVal()`), updating that value upon a suitable `observeEvent()` event with relevant `customdata`, and clearing that reactive value (i.e., `cars(NULL)`) in response to another event is a very useful pattern to support essentially any sort of linked views paradigm because the logic behind the resolution of selection sequences is under your complete control in R. For example, Figure 17.14 simply accumulates the event data from `"plotly_hover"` (which is like a logical OR operations), but for other applications, you may need different logic, like the AND, XOR, etc.

```
plotly_example("shiny", "event_data_persist")
```

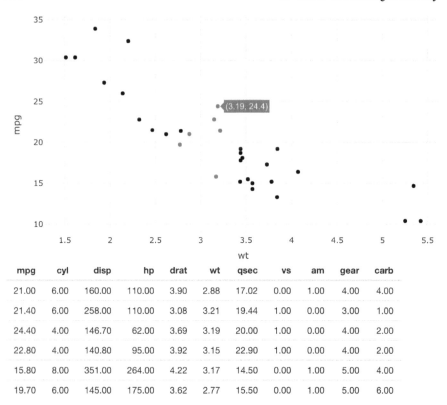

mpg	cyl	disp	hp	drat	wt	qsec	vs	am	gear	carb
21.00	6.00	160.00	110.00	3.90	2.88	17.02	0.00	1.00	4.00	4.00
21.40	6.00	258.00	110.00	3.08	3.21	19.44	1.00	0.00	3.00	1.00
24.40	4.00	146.70	62.00	3.69	3.19	20.00	1.00	0.00	4.00	2.00
22.80	4.00	140.80	95.00	3.92	3.15	22.90	1.00	0.00	4.00	2.00
15.80	8.00	351.00	264.00	4.22	3.17	14.50	0.00	1.00	5.00	4.00
19.70	6.00	145.00	175.00	3.62	2.77	15.50	0.00	1.00	5.00	6.00

FIGURE 17.14: Using `reactiveVals()` to enable a persistent brush via mouse hover. In this example, the brush can be cleared through a double-click event. For a video demonstration of the interactive, see `https://bit.ly/shiny-hover-persist`. For the interactive, see `https://plotly-r.com/interactives/shiny-hover-persist.html`

Figure 17.15 demonstrates a **shiny** gadget for interactively remov-ing/adding points from a linear model via a scatterplot. A **shiny** gadget is similar to a normal **shiny** app except that it allows you to return object(s) from the application back to into your R session. In this case, Figure 17.15 returns the fitted model with the outliers removed and the chosen polynomial degree. The logic behind this app does more than simply accumulate event data every time a point is clicked. Instead, it adds points to the 'outlier' set only if it isn't already an outlier, and removes points that are already in the "outlier" set (so, it's essentially XOR logic).

```
plotly_example("shiny", "lmGadget")
```

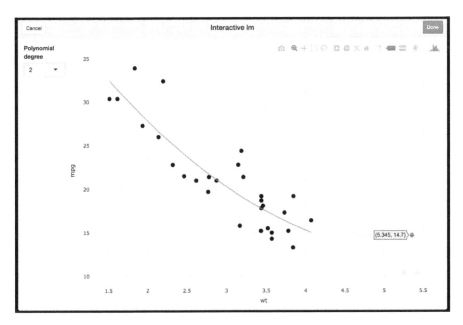

FIGURE 17.15: Interactively removing observations from a linear model. Credit to Winston Chang for the initial implementation of this **shiny** gadget using `shiny::plotOutput()` instead of `plotly::plotlyOutput()`. For a video demonstration of the interactive, see `https://bit.ly/shiny-lmGadget`. For the interactive, see `https://plotly-r.com/interactives/shiny-lmGadget.html`

As you can already see, the ability to accumulate and manage event data is a critical skill to have in order to implement **shiny** applications with complex interactive capabilities. The pattern demonstrated here is known more generally as "maintaining state" of a **shiny** app based on user interactions and has a variety of applications. So far, we've really only seen how to maintain state of a single view, but as we'll see later in Section 17.4, the ability to maintain state is required to implement many advanced applications of multiple linked views. Also, it should be noted that Figures 17.14 and 17.15 perform a full redraw when updated; these apps would feel a bit more responsive if they leveraged strategies from Section 17.3.1.

17.3   Improving performance

Multiple linked views are known to help facilitate data exploration, but latency in the user interface is also known to reduce exploratory findings (Liu and Heer, 2014). In addition to the advice and techniques offered in Section 17.3.1 for improving **plotly**'s performance in general, there are also techniques specifically for **shiny** apps that you can leverage to help improve the user experience.

When trying to speed up any slow code, the first step is always to identify the main contributor(s) to the poor performance. In some cases, your intuition may serve as a helpful guide, but in order to *really* see what's going on, consider using a code profiling tool like **profvis** (Chang and Luraschi, 2018). The **profvis** package provides a really nice way to visualize and isolate slow running R code in general, but it also works well for profiling **shiny** apps (RStudio, 2014b).

A lot of different factors can contribute to poor performance in a **shiny** app, but thankfully, the **shiny** ecosystem provides an extensive toolbox for diagnosing and improving performance. The **profvis** package is great for identifying "universal" performance issues, but when deploying shiny apps into production, there may be other potential bottlenecks that surface. This is largely due to R's single-threaded nature – a single R server has difficulty scaling to many users because, by default, it can only handle one job at a time. The **shinyloadtest** package helps to identify those bottlenecks and **shiny**'s support for asynchronous programming with **promises** is one way to address them without increasing computational infrastructure (e.g., multiple servers) (Dipert et al., 2018; Cheng, 2018b).

To reiterate the section on "Improving performance and scalability" in **shiny** from Cheng (2018a), you have a number of tools available to address performance:

1. The **profvis** package for profiling code.
2. Cache computations ahead-of-time.
3. Cache computations at run time.

4. Cache computations through chaining reactive expressions.
5. Leverage multiple R processes and/or servers.
6. Async programming with **promises**.

We won't directly cover these topics, but it's worth noting that all these tools are primarily designed for improving *server-side* performance of a **shiny** app. It could be that sluggish plots in your **shiny** app are due to sluggish server-side code, but it could also be that some of the sluggishness is due to redundant work being done client-side by **plotly**. Avoiding this redundancy, as covered in Section 17.3.1, can be difficult, and it doesn't always lead to noticeable improvements. However, when you need to put lots of graphical elements on a plot, then update just a portion of the plot in response to user event(s), the added complexity can be worth the effort.

17.3.1   Partial plotly updates

By default, when `renderPlotly()` renders a new **plotly** graph, it's essentially equivalent to executing a block of R code from your R prompt and generating a new **plotly** graph from scratch. That means, not only does the R code need to re-execute to generate a new R object, but it also has to re-serialize that object as JSON, and your browser has to re-render the graph from the new JSON object (more on this in Chapter 24). In cases where your **plotly** graph does not need to serialize a lot of data and/or render lots of graphical elements, as in Figure 17.1, you can likely perform a full redraw without noticeable glitches, especially if you use Canvas-based rendering rather than SVG (i.e., `toWebGL()`). Generally speaking, you should try very hard to make your app responsive before adopting partial **plotly** updates in **shiny**. It makes your app logic easy to reason about because you don't have to worry about maintaining the state of the graph, but sometimes you have no other choice.

On initial page load, **plotly** graphs must be drawn from stratch, but when responding to certain user events, often a partial update to an existing plot is sufficient and more responsive. Take, for instance, the difference between Figure 17.16, which does a full redraw on every up-

date, and Figure 17.17, which does a partial update after initial load.
Both of these **shiny** apps display a scatterplot with 100,000 points and
allow a user to overlay a fitted line through a checkbox. The key difference is that in Figure 17.16, the **plotly** graph is regenerated from scratch
every time the value of input$smooth changes; whereas in Figure 17.17
only the fitted line is added/removed from the **plotly**. Since the main
bottleneck lies in redrawing the points, Figure 17.17 can add/remove
the fitted line in a much more responsive fashion.

```r
library(shiny)
library(plotly)

# Generate 100,000 observations from 2 correlated random variables
s <- matrix(c(1, 0.5, 0.5, 1), 2, 2)
d <- MASS::mvrnorm(1e6, mu = c(0, 0), Sigma = s)
d <- setNames(as.data.frame(d), c("x", "y"))

# fit a simple linear model
m <- lm(y ~ x, data = d)

# generate y predictions over a grid of 10 x values
dpred <- data.frame(
  x = seq(min(d$x), max(d$x), length.out = 10)
)
dpred$yhat <- predict(m, newdata = dpred)

ui <- fluidPage(
  plotlyOutput("scatterplot"),
  checkboxInput(
    "smooth",
    label = "Overlay fitted line?",
    value = FALSE
  )
)

server <- function(input, output, session) {
```

```
output$scatterplot <- renderPlotly({

  p <- plot_ly(d, x   = ~x, y = ~y) %>%
    add_markers(color = I("black"), alpha = 0.05) %>%
    toWebGL() %>%
    layout(showlegend = FALSE)

  if (!input$smooth) return(p)

  add_lines(p, data = dpred, x = ~x, y = ~yhat, color = I("red"))
  })

}

shinyApp(ui, server)
```

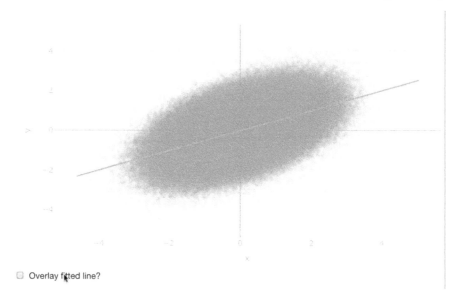

☐ Overlay fitted line?

FIGURE 17.16: Naive implementation of a shiny app that optionally overlays a fitted line to a scatterplot. A full redraw of the plot is performed every time the checkbox is clicked, leading to an unnecessarily slow plot. For a video demonstration of the interactive, see `https://bit.ly/shiny-scatterplot`. For the interactive, see `https://plotly-r.com/interactives/shiny-scatterplot.html`

In terms of the implementation behind Figures 17.16 and 17.17, the only difference resides in the `server` definition. In Figure 17.17, the `renderPlotly()` statement no longer has a dependency on input values, so that code is only executed once (on page load) to generate the initial view of the scatterplot. The logic behind adding and removing the fitted line is handled through an `observe()` block; this reactive expression watches the `input$smooth` input value and modifies the `output$scatterplot` widget whenever it changes. To trigger a modification of a **plotly** output widget, you must create a proxy object with `plotlyProxy()` that references the relevant output ID. Once a proxy object is created, you can invoke any sequence of plotly.js function(s)[6] on it with `plotlyProxyInvoke()`. Invoking a method with the correct arguments can be tricky and requires knowledge of plotly.js because

---

[6]`https://plot.ly/javascript/plotlyjs-function-reference/#plotlymaketemplate`

`plotlyProxyInvoke()` will send these arguments directly to the plotly.js method and therefore doesn't support the same 'high-level' semantics that `plot_ly()` does.

```r
server <- function(input, output, session) {

  output$scatterplot <- renderPlotly({
    plot_ly(d, x = ~x, y = ~y) %>%
      add_markers(color = I("black"), alpha = 0.05) %>%
      toWebGL()
  })

  observe({
    if (input$smooth) {
      # this is essentially the plotly.js way of doing
      # `p %>% add_lines(x = ~x, y = ~yhat) %>% toWebGL()`
      # without having to redraw the entire plot
      plotlyProxy("scatterplot", session) %>%
        plotlyProxyInvoke(
          "addTraces",
          list(
            x = dpred$x,
            y = dpred$yhat,
            type = "scattergl",
            mode = "lines",
            line = list(color = "red")
          )
        )
    } else {
      # JavaScript index starts at 0, so the '1' here really means
      # "delete the second traces (i.e., the fitted line)"
      plotlyProxy("scatterplot", session) %>%
        plotlyProxyInvoke("deleteTraces", 1)
    }
  })
}
```

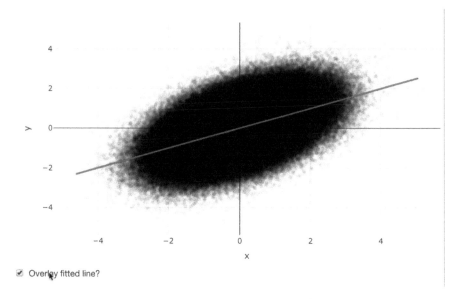

FIGURE 17.17: A more responsive version of Figure 17.16. For a video demonstration of the interactive, see `https://bit.ly/shiny-scatterplot-performant`. For the interactive, see `https://plotly-r.com/interactives/shiny-scatterplot-performant.html`

Figure 17.16 demonstrates a common use case where partial updates can be helpful, but there are other not-so-obvious cases. The next section covers a range of examples where you'll see how to leverage partial updates to implement smooth 'streaming' visuals, avoid resetting axis ranges, avoid flickering basemap layers, and more.

17.3.2   Partial update examples

The last section explains why you may want to leverage partial **plotly** updates in **shiny** to get more responsive updates through an example. That example leveraged the plotly.js functions[7] `Plotly.addTraces()` and `Plotly.deleteTraces()` to add/remove a layer to a plot after its initial draw. There are numerous other plotly.js functions that can be handy for a variety of use cases, some of the most widely used ones are: `Plotly.restyle()` for updating data visuals (Section

---

[7] `https://plot.ly/javascript/plotlyjs-function-reference/`

17.3.2.1), `Plotly.relayout()` for updating the layout (Section 17.3.2.2), and `Plotly.extendTraces()` for streaming data (Section 17.3.2.3).

17.3.2.1 Modifying traces

All **plotly** figures have two main components: traces (i.e., mapping from data to visuals) and layout. The plotly.js function `Plotly.restyle()` is for modifying any existing traces. In addition to being a performant way to modify existing data and/or visual properties, it also has the added benefit of not affecting the current layout of the graph. Notice how, in Figure 17.18 for example, when the size of the marker/path changes, it doesn't change the camera's view of the 3D plot that the user altered after initial draw. If these input widgets triggered a full redraw of the plot, the camera would be reset to its initial state.

```
plotly_example("shiny", "proxy_restyle_economics")
```

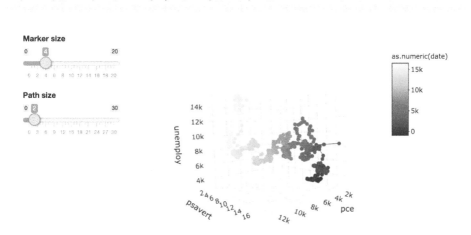

FIGURE 17.18: Using `Plotly.restyle()` to change just the width of a path and markers along that path in response to changes to **shiny** input sliders. For a video demonstration of the interactive, see https://bit.ly/shiny-partial-restyle. For the interactive, see https://plotly-r.com/interactives/shiny-partial-restyle.html

One un-intuitive thing about `Plotly.restyle()` is that it fully replaces object (i.e., attributes that contain attributes) definitions like `marker` by default. To modify just a particular attribute of an object, like the size of a marker, you must replace that attribute directly (hence `marker.size`). As mentioned in the official documentation[8], by default, modifications are applied to all traces, but specific traces can be targeted through their trace index (which starts at 0, because of JavaScript)!

17.3.2.2   Updating the layout

All **plotly** figures have two main components: traces (i.e., mapping from data to visuals) and layout. The plotly.js function `Plotly.relayout()` modifies the layout component, so it can control a wide variety of things such as titles, axis definitions, annotations, shapes, and many other things. It can even be used to change the basemap layer of a Mapbox-powered layout, as in Figure 17.19. Note how this example uses `schema()` to grab all the pre-packaged basemap layers and create a dropdown of those options, but you can also provide a URL to a custom basemap style[9].

```r
plotly_example("shiny", "proxy_mapbox")
```

---

[8]https://plot.ly/javascript/plotlyjs-function-reference/#plotlyrestyle
[9]https://www.mapbox.com/help/create-a-custom-style/

FIGURE 17.19: Using a `shiny::selectInput()` to modify the basemap style of `plot_mapbox()` via `Plotly.relayout()`. For a video demonstration of the interactive, see `https://bit.ly/shiny-mapbox-relayout`. For the interactive, see `https://plotly-r.com/interactives/shiny-mapbox-relayout.html`

Figure 17.20 demonstrates a clever use of `Plotly.relayout()` to set the y-axis range in response to changes in the x-axis range.

```
plotly_example("shiny", "proxy_relayout")
```

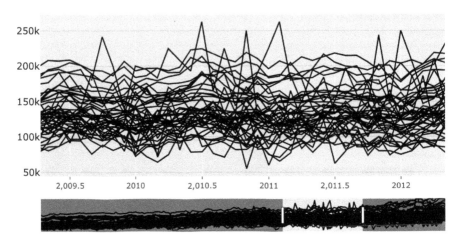

FIGURE 17.20: Using `Plotly.relayout()` to 'auto-range' the y-axis in response to changes in the x-axis range. For a video demonstration of the interactive, see `https://bit.ly/shiny-rangeslider-relayout`. For the interactive, see `https://plotly-r.com/interactives/shiny-rangeslider-relayout.html`

17.3.2.3 Streaming data

At this point, we've seen how to add/remove traces (e.g., add/remove a fitted line, as in Figure 17.17), and how to edit specific trace properties (e.g., change marker size or path width, as in Figure 17.18), but what about adding more data to existing trace(s)? This is a job for the plotly.js function `Plotly.extendTraces()` and/or `Plotly.prependTraces()` which can be used to efficiently 'stream' data into an existing plot, as done in Figure 17.21.

The implementation behind Figure 17.21, an elementary example of a random walk, makes use of some fairly sophisticated reactive programming tools from **shiny**. Similar to most examples from this section, the `renderPlotly()` statement is executed once on initial load to draw the initial line with just two data points. By default, the plot is not streaming, but streaming can be turned on or off through the click of a button, which will require the app to know (at all times) whether or not we are in a streaming state. One way to do this is to leverage **shiny**'s `reactiveValues()`, which act like input values, but can be created and modified entirely server-side, making them quite useful for maintaining the state of an application. In this case, the reactive value rv$stream is used to store the streaming state, which is turned on/off whenever the actionButton() is clicked (via the observeEvent() logic).

Even if the app is not streaming, there is still constant client/server communication because of the use of invalidateLater() inside the observe(). This effectively tells **shiny** to re-evaluate the observe() block every 100 milliseconds. If the app isn't in streaming mode, then it exits early without doing anything. If the app is streaming, then we first use sample() to randomly draw either -1 or 1 (with equal probability) and use the result to update the most recent (x, y) state. This is done by assigning a new value to the reactive values rv$y and rv$n within an isolate()d context. If this assignment happened outside of an isolate()d context, it would cause the reactive expression to be invalidated and cause an infinite loop! Once we have the new (x, y) point stored away, Plotly.extendTraces() can be used to add the new point to the plotly graph.

```
plotly_example("shiny", "stream")
```

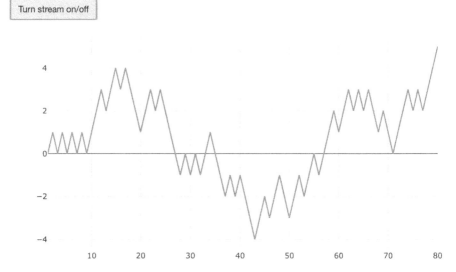

FIGURE 17.21: Using `Plotly.extendTraces()` to efficiently stream data into a plotly chart. This specific example implements a random walk (using R's random number generator) which updates every 100 milliseconds. For a video demonstration of the interactive, see `https://bit.ly/shiny-stream`. For the interactive, see `https://plotly-r.com/interactives/shiny-stream.html`

To see more examples that leverage partial updating, see Section 17.4.2.

17.4   Advanced applications

This section combines concepts from prior sections in linking views with shiny and applies them towards some popular use cases.

17.4.1   Drill-down

Figure 17.22 displays sales broken down by business category (e.g., Furniture, Office Supplies, Technology) in a pie chart. It allows the user to click on a slice of the pie to 'drill-down' into sub-categories of the chosen category. In terms of the implementation, the key aspect here is to maintain the state of the currently selected category via a reactiveVal() (see more in Section 17.2.8) and update that value when either a category is clicked or the "Back" button is pressed. This may seem like a lot of code to get a basic drill-down pie chart, but the core reactivity concepts in this implementation translate well to more complex drill-down applications.

```r
library(shiny)
library(dplyr)
library(readr)
library(purrr) # just for `%||%`

sales <- read_csv("https://plotly-r.com/data-raw/sales.csv")
categories <- unique(sales$category)

ui <- fluidPage(plotlyOutput("pie"), uiOutput("back"))

server <- function(input, output, session) {
  # for maintaining the current category (i.e., selection)
  current_category <- reactiveVal()

  # report sales by category, unless a category is chosen
  sales_data <- reactive({
    if (!length(current_category())) {
      return(count(sales, category, wt = sales))
    }
    sales %>%
      filter(category %in% current_category()) %>%
      count(sub_category, wt = sales)
  })

  # Note that pie charts don't currently attach the label/value
  # with the click data, but we can include as `customdata`
  output$pie <- renderPlotly({
    d <- setNames(sales_data(), c("labels", "values"))
    plot_ly(d) %>%
```

```r
    add_pie(
      labels = ~labels,
      values = ~values,
      customdata = ~labels
    ) %>%
    layout(title = current_category() %||% "Total Sales")
  })

  # update the current category when appropriate
  observe({
    cd <- event_data("plotly_click")$customdata[[1]]
    if (isTRUE(cd %in% categories)) current_category(cd)
  })

  # populate back button if category is chosen
  output$back <- renderUI({
    if (length(current_category()))
      actionButton("clear", "Back", icon("chevron-left"))
  })

  # clear the chosen category on back button press
  observeEvent(input$clear, current_category(NULL))
}

shinyApp(ui, server)
```

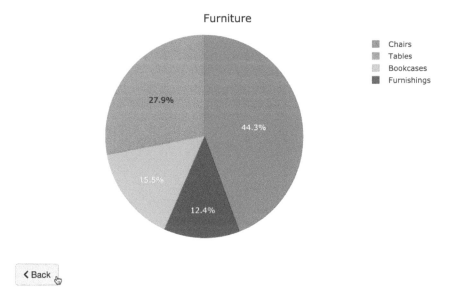

FIGURE 17.22: A drill-down pie chart of sales by category. By clicking on a category (e.g., Furniture), the pie chart updates to display sales by sub-categories within Furniture. For a video demonstration of the interactive, see `https://bit.ly/shiny-drill-down-pie`. For the interactive, see `https://plotly-r.com/interactives/shiny-drill-down-pie.html`

A basic drill-down like Figure 17.22 is somewhat useful on its own, but it becomes much more useful when linked to multiple views of the data. Figure 17.23 improves on Figure 17.22 to show sales over time by the category or sub-category (if a category is currently chosen). Note that the key aspect of implementation remains the same (i.e., maintaining state via `reactiveValue()`); the main difference is that the time series view now also responds to changes in the currently selected category. That is, both views show sales by category when no category is selected, and sales by sub-category when a category is selected.

```
library(shiny)
library(dplyr)
library(readr)
```

```r
sales <- read_csv("https://plotly-r.com/data-raw/sales.csv")
categories <- unique(sales$category)

ui <- fluidPage(
  plotlyOutput("bar"),
  uiOutput("back"),
  plotlyOutput("time")
)

server <- function(input, output, session) {

  current_category <- reactiveVal()

  # report sales by category, unless a category is chosen
  sales_data <- reactive({
    if (!length(current_category())) {
      return(count(sales, category, wt = sales))
    }
    sales %>%
      filter(category %in% current_category()) %>%
      count(sub_category, wt = sales)
  })

  # the pie chart
  output$bar <- renderPlotly({
    d <- setNames(sales_data(), c("x", "y"))

    plot_ly(d) %>%
      add_bars(x = ~x, y = ~y, color = ~x) %>%
      layout(title = current_category() %||% "Total Sales")
  })

  # same as sales_data
  sales_data_time <- reactive({
    if (!length(current_category())) {
      return(count(sales, category, order_date, wt = sales))
    }
```

```
    sales %>%
      filter(category %in% current_category()) %>%
      count(sub_category, order_date, wt = sales)
  })

  output$time <- renderPlotly({
    d <- setNames(sales_data_time(), c("color", "x", "y"))
    plot_ly(d) %>%
      add_lines(x = ~x, y = ~y, color = ~color)
  })

  # update the current category when appropriate
  observe({
    cd <- event_data("plotly_click")$x
    if (isTRUE(cd %in% categories)) current_category(cd)
  })

  # populate back button if category is chosen
  output$back <- renderUI({
    if (length(current_category()))
      actionButton("clear", "Back", icon("chevron-left"))
  })

  # clear the chosen category on back button press
  observeEvent(input$clear, current_category(NULL))
}

shinyApp(ui, server)
```

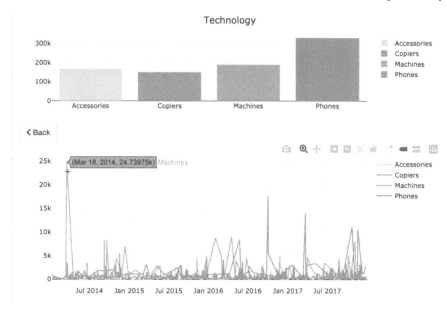

FIGURE 17.23: Coordinating drill-down across multiple views. By clicking on a category (e.g., Technology), both the bar chart and the time-series updates to display sales within Technology. For a video demonstration of the interactive, see `https://bit.ly/shiny-drill-down-bar-time`. For the interactive, see `https://plotly-r.com/interactives/shiny-drill-down-bar-time.html`

Figures 17.22 and 17.23 demonstrate one level of drill-down; what about multiple levels? Introducing multiple levels adds complexity not only to the implementation, but also the user experience. Especially in a drill-down approach where the *same view* is being filtered, it can be difficult for users to remember how they got to a particular view. Figure 17.24 demonstrates how we could extend Figure 17.23 to implement yet another level of drilling down (i.e., category -> sub-category -> product IDs) as well as populate a `selectInput()` dropdown for each active drill-down category. Not only does this help users to remember how they got to the particular view, but it also provides the ability to easily modify the sequence of drill-down events. In terms of implementation, the main idea is very similar to before: we still store and update the state of each 'drill-down' in its own reactive value, but now when a 'parent' category changes (e.g., category) it should invalidate (i.e., clear) any currently selected 'child' categories (e.g., sub_category).

```r
library(shiny)
library(plotly)
library(dplyr)
library(readr)

sales <- read_csv("https://plotly-r.com/data-raw/sales.csv")
categories <- unique(sales$category)
sub_categories <- unique(sales$sub_category)
ids <- unique(sales$id)

ui <- fluidPage(
  uiOutput("history"),
  plotlyOutput("bars", height = 200),
  plotlyOutput("lines", height = 300)
)

server <- function(input, output, session) {
  # These reactive values keep track of the drilldown state
  # (NULL means inactive)
  drills <- reactiveValues(
    category = NULL,
    sub_category = NULL,
    id = NULL
  )
  # filter the databased on active drill-downs
  # also create a column, value, which keeps track of which
  # variable we're interested in
  sales_data <- reactive({
    if (!length(drills$category)) {
      return(mutate(sales, value = category))
    }
    sales <- filter(sales, category %in% drills$category)
```

```r
  if (!length(drills$sub_category)) {
    return(mutate(sales, value = sub_category))
  }
  sales <- filter(sales, sub_category %in% drills$sub_category)
  mutate(sales, value = id)
})


# bar chart of sales by 'current level of category'
output$bars <- renderPlotly({
  d <- count(sales_data(), value, wt = sales)

  p <- plot_ly(d, x = ~value, y = ~n, source = "bars") %>%
    layout(
      yaxis = list(title = "Total Sales"),
      xaxis = list(title = "")
    )

  if (!length(drills$sub_category)) {
    add_bars(p, color = ~value)
  } else if (!length(drills$id)) {
    add_bars(p) %>%
      layout(
        hovermode = "x",
        xaxis = list(showticklabels = FALSE)
      )
  } else {
    # add a visual cue of which ID is selected
    add_bars(p) %>%
      filter(value %in% drills$id) %>%
      add_bars(color = I("black")) %>%
      layout(
        hovermode = "x", xaxis = list(showticklabels = FALSE),
        showlegend = FALSE, barmode = "overlay"
      )
  }
})
```

```r
# time-series chart of the sales
output$lines <- renderPlotly({
  p <- if (!length(drills$sub_category)) {
    sales_data() %>%
      count(order_date, value, wt = sales) %>%
      plot_ly(x = ~order_date, y = ~n) %>%
      add_lines(color = ~value)
  } else if (!length(drills$id)) {
    sales_data() %>%
      count(order_date, wt = sales) %>%
      plot_ly(x = ~order_date, y = ~n) %>%
      add_lines()
  } else {
    sales_data() %>%
      filter(id %in% drills$id) %>%
      select(-value) %>%
      plot_ly() %>%
      add_table()
  }
  p %>%
    layout(
      yaxis = list(title = "Total Sales"),
      xaxis = list(title = "")
    )
})


# control the state of the drilldown by clicking the bar graph
observeEvent(event_data("plotly_click", source = "bars"), {
  x <- event_data("plotly_click", source = "bars")$x
  if (!length(x)) return()

  if (!length(drills$category)) {
    drills$category <- x
  } else if (!length(drills$sub_category)) {
    drills$sub_category <- x
  } else {
    drills$id <- x
```

```
    }
  })


  # populate a `selectInput()` for each active drilldown
  output$history <- renderUI({
    if (!length(drills$category))
      return("Click the bar chart to drilldown")

    categoryInput <- selectInput(
      "category", "Category",
      choices = categories, selected = drills$category
    )
    if (!length(drills$sub_category)) return(categoryInput)
    sd <- filter(sales, category %in% drills$category)
    subCategoryInput <- selectInput(
      "sub_category", "Sub-category",
      choices = unique(sd$sub_category),
      selected = drills$sub_category
    )
    if (!length(drills$id)) {
      return(fluidRow(
        column(3, categoryInput),
        column(3, subCategoryInput)
      ))
    }
    sd <- filter(sd, sub_category %in% drills$sub_category)
    idInput <- selectInput(
      "id", "Product ID",
      choices = unique(sd$id), selected = drills$id
    )
    fluidRow(
      column(3, categoryInput),
      column(3, subCategoryInput),
      column(3, idInput)
    )
  })
```

```
  # control the state of the drilldown via the `selectInput()`s
  observeEvent(input$category, {
    drills$category <- input$category
    drills$sub_category <- NULL
    drills$id <- NULL
  })
  observeEvent(input$sub_category, {
    drills$sub_category <- input$sub_category
    drills$id <- NULL
  })
  observeEvent(input$id, {
    drills$id <- input$id
  })
}


shinyApp(ui, server)
```

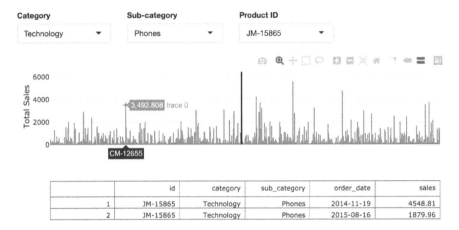

FIGURE 17.24: Coordinating drill-down across multiple views. By clicking on a category (e.g., Technology), both the bar chart and the time-series updates to display sales within Technology. For a video demonstration of the interactive, see `https://bit.ly/shiny-drill-down-bar-time`. For the interactive, see `https://plotly-r.com/interactives/shiny-drill-down-bar-time.html`

Another way to make it easier for the user to recall their drill-down sequence is to generate a *new* view based on the selection. Figure 17.25 allows one to click on a category (e.g., Furniture) to generate another bar chart of sales broken down by that category's sub-categories (e.g., Bookcases, Chairs, etc.). From there, a sub-category may be clicked to populate a time series of sales for that sub-category. Finally, by clicking on the time series, a table of sales from that date are displayed.

Similar to Figure 17.24, changes at a given category level cause invalidation of all child categories (in this case, all downstream views are cleared). For example, note how in Figure 17.25, a click of a category clears the sub-category and order-date. Moreover, a change in `sub_category` clears the `order_date`, but effect the current `category`.

```
library(shiny)
library(plotly)
library(dplyr)
```

```r
library(readr)

sales <- read_csv("https://plotly-r.com/data-raw/sales.csv")

ui <- fluidPage(
  plotlyOutput("category", height = 200),
  plotlyOutput("sub_category", height = 200),
  plotlyOutput("sales", height = 300),
  dataTableOutput("datatable")
)

# avoid repeating this code
axis_titles <- . %>%
  layout(
    xaxis = list(title = ""),
    yaxis = list(title = "Sales")
  )

server <- function(input, output, session) {

  # for maintaining the state of drill-down variables
  category <- reactiveVal()
  sub_category <- reactiveVal()
  order_date <- reactiveVal()

  # when clicking on a category,
  observeEvent(event_data("plotly_click", source = "category"), {
    category(event_data("plotly_click", source = "category")$x)
    sub_category(NULL)
    order_date(NULL)
  })

  observeEvent(event_data("plotly_click", source = "sub_category"), {
    sub_category(
      event_data("plotly_click", source = "sub_category")$x
    )
    order_date(NULL)
```

```r
})

observeEvent(event_data("plotly_click", source = "order_date"), {
  order_date(event_data("plotly_click", source = "order_date")$x)
})

output$category <- renderPlotly({
  sales %>%
    count(category, wt = sales) %>%
    plot_ly(x = ~category, y = ~n, source = "category") %>%
    axis_titles() %>%
    layout(title = "Sales by category")
})

output$sub_category <- renderPlotly({
  if (is.null(category())) return(NULL)

  sales %>%
    filter(category %in% category()) %>%
    count(sub_category, wt = sales) %>%
    plot_ly(x = ~sub_category, y = ~n, source = "sub_category") %>%
    axis_titles() %>%
    layout(title = category())
})

output$sales <- renderPlotly({
  if (is.null(sub_category())) return(NULL)

  sales %>%
    filter(sub_category %in% sub_category()) %>%
    count(order_date, wt = sales) %>%
    plot_ly(x = ~order_date, y = ~n, source = "order_date") %>%
    add_lines() %>%
    axis_titles() %>%
    layout(title = paste(sub_category(), "sales over time"))
})
```

```r
output$datatable <- renderDataTable({
  if (is.null(order_date())) return(NULL)

  sales %>%
    filter(
      sub_category %in% sub_category(),
      as.Date(order_date) %in% as.Date(order_date())
    )
})

}

shinyApp(ui, server)
```

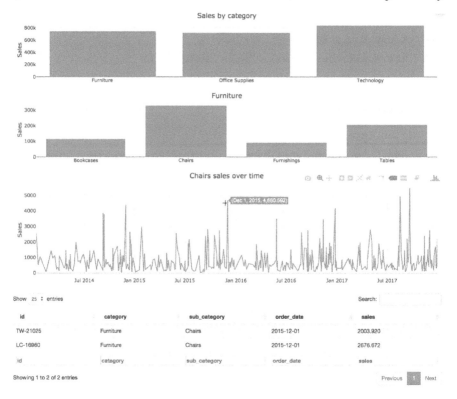

FIGURE 17.25: Using a drill-down approach to navigating through sales data by category, sub-category, and order date. For a video demonstration of the interactive, see `https://bit.ly/shiny-drill-down`. For the interactive, see `https://plotly-r.com/interactives/shiny-drill-down.html`

17.4.2 Cross-filter

Somewhat related to the drill-down idea is the so-called 'cross-filter' chart. The main difference between drill-down and cross-filter is that, with cross-filter, interactions don't generate new charts; interactions impose a filter on the data shown in a fixed set of multiple views. The typical cross-filter implementation allows for multiple brushes (i.e., filters) and uses the intersection of those filters to the dataset displayed in those views. Implementing a scalable and responsive cross-filter with three or more views can get quite complicated; we'll walk through some simple examples first for learning purposes, then progress to more sophisticated and complex applications.

Figure 17.26 demonstrates the simplest way to implement a cross-filter between *two* histograms. It uses the arrival (arr_time) and departure (dep_time) from the flights dataset via the **nycflights13** package (Wickham, 2018a). Notice how, in the implementation, the dep_time view is re-drawn from stratch every time the arr_time brush changes (and vice versa). Not only is it completely redrawn (i.e., it relies on renderPlotly() to perform the update), but it also uses add_histogram() which performs binning client-side (in the web browser). That means, every time a brush changes, the **shiny** server sends *all the raw data* to the browser and plotly.js redraws the histogram from scratch.

```r
library(shiny)
library(dplyr)
library(nycflights13)

ui <- fluidPage(
  plotlyOutput("arr_time"),
  plotlyOutput("dep_time")
)

server <- function(input, output, session) {

  output$arr_time <- renderPlotly({
    p <- plot_ly(flights, x = ~arr_time, source = "arr_time")

    brush <- event_data("plotly_brushing", source = "dep_time")
    if (is.null(brush)) return(p)

    p %>%
      filter(between(dep_time, brush$x[1], brush$x[2])) %>%
      add_histogram()
  })

  output$dep_time <- renderPlotly({
    p <- plot_ly(flights, x = ~dep_time, source = "dep_time")
```

```
brush <- event_data("plotly_brushing", source = "arr_time")
if (is.null(brush)) return(p)

p %>%
  filter(between(arr_time, brush$x[1], brush$x[2])) %>%
  add_histogram()
})


}


shinyApp(ui, server)
```

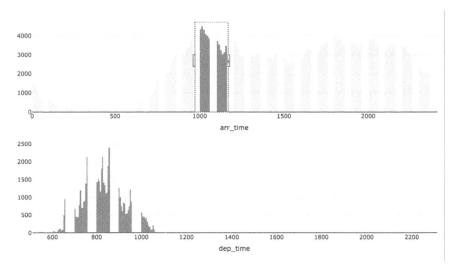

FIGURE 17.26: A 'naive' cross-filter implementation linking arrival time with departure time of about 350,000 `flights` from the **nycflights13** package. For a video demonstration of the interactive, see https://bit. ly/shiny-crossfilter-preview. For the interactive, see https://plotly-r.com/interactives/shiny-crossfilter-naive.html

Although the video behind Figure 17.26 demonstrates the app is fairly responsive at 350,000 observations, this implementation won't scale to much larger data, especially if being viewed with a poor internet connection. I call this a 'naive' implementation because the reactive logic is easy to reason about, but it illustrates two common issues that we can address to gain speed improvements:

1. More data than necessary being sent 'over-the-wire' (i.e., between the server and the client). This idea is not unique to **shiny** applications; with any web application framework, it's important to minimize the amount of data your requesting from a server if you want a responsive website.
2. More client-side rendering work than necessary to achieve the request update.

The implementation behind Figure 17.26 could improve in these areas by doing the following:

1. Perform the binning server-side instead of client-side. This will reduce the amount of data needed to be sent from server to client so that responsiveness is less dependent on a good internet connection. Here we propose using **ggstat** for the server-side binning since it's fairly fast and simple if you're data can fit into memory (Wickham, 2016). If your data does not fit into memory you could use something like **dbplot** to perform the binning in a database (Ruiz, 2018).
2. Perform less rendering work client-side. That is, instead of relying on `renderPlotly()` to re-render the chart from scratch every time the charts need an update, we could instead modify just the bar heights (using the techniques from Section 17.3.1).

```
library(shiny)
library(dplyr)
library(nycflights13)
library(ggstat)

arr_time <- flights$arr_time
```

```r
dep_time <- flights$dep_time
arr_bins <- bin_fixed(arr_time, bins = 250)
dep_bins <- bin_fixed(dep_time, bins = 250)
arr_stats <- compute_stat(arr_bins, arr_time) %>%
  filter(!is.na(xmin_))
dep_stats <- compute_stat(dep_bins, dep_time) %>%
  filter(!is.na(xmin_))

ui <- fluidPage(
  plotlyOutput("arr_time", height = 250),
  plotlyOutput("dep_time", height = 250)
)

server <- function(input, output, session) {

  output$arr_time <- renderPlotly({
    plot_ly(arr_stats, source = "arr_time") %>%
      add_bars(x = ~xmin_, y = ~count_)
  })

  output$dep_time <- renderPlotly({
    plot_ly(dep_stats, source = "dep_time") %>%
      add_bars(x = ~xmin_, y = ~count_)
  })

  # arr_time brush updates dep_time view
  observe({
    brush <- event_data("plotly_brushing", source = "arr_time")
    p <- plotlyProxy("dep_time", session)

    # if brush is empty, restore default
    if (is.null(brush)) {
      plotlyProxyInvoke(p, "restyle", "y", list(dep_stats$count_), 0)
    } else {
      in_filter <- between(dep_time, brush$x[1], brush$x[2])
      dep_count <- dep_bins %>%
        compute_stat(dep_time[in_filter]) %>%
```

```
      filter(!is.na(xmin_)) %>%
      pull(count_)

    plotlyProxyInvoke(p, "restyle", "y", list(dep_count), 0)
  }
})


observe({
  brush <- event_data("plotly_brushing", source = "dep_time")
  p <- plotlyProxy("arr_time", session)

  # if brush is empty, restore default
  if (is.null(brush)) {
    plotlyProxyInvoke(p, "restyle", "y", list(arr_stats$count_), 0)
  } else {
    in_filter <- between(arr_time, brush$x[1], brush$x[2])
    arr_count <- arr_bins %>%
      compute_stat(arr_time[in_filter]) %>%
      filter(!is.na(xmin_)) %>%
      pull(count_)

    plotlyProxyInvoke(p, "restyle", "y", list(arr_count), 0)
  }
})


}


shinyApp(ui, server)
```

Before we address the additional complexity that comes with linking three or more views, let's consider targeting a 2D distribution in the cross-filter, as in Figure 17.27. This approach uses kde2d() from the

MASS package to summarize the 2D distribution server-side rather than attempting to show all the raw data in a scatterplot.[10]

```
plotly_example("shiny", "crossfilter_kde")
```

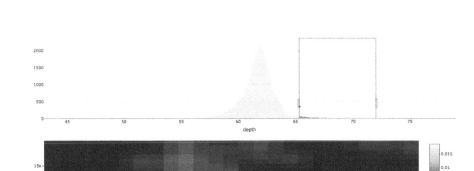

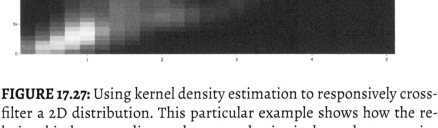

FIGURE 17.27: Using kernel density estimation to responsively cross-filter a 2D distribution. This particular example shows how the relationship between diamond carat and price is dependent upon its depth. For a video demonstration of the interactive, see https://bit.ly/shiny-crossfilter-kde. For the interactive, see https://plotly-r.com/interactives/shiny-crossfilter-kde.html

When linking three or more views in a cross-filter, it's important to have a mechanism to maintain the state of all the active brushes. This is because, when updating a given view, it needs to know about *all* of the active brushes. The implementation behind Figure 17.28 maintains the range of every active brush through a `reactiveValues()` variable named `brush_ranges`. Every time a brush changes, the state of `brush_ranges` is updated, then it is used to filter the data down to

---

[10] It's possible to do this responsively with about 50,000 observations, but it won't scale to anything much larger than that. Run `plotly_example("shiny", "crossfilter_scatter")` to see it in action as well as a corresponding video at https://vimeo.com/318129005

the relevant observations. That filtered data is then binned and used to modify the bar heights of every view (except for the one being brushed).

```
plotly_example("shiny", "crossfilter")
```

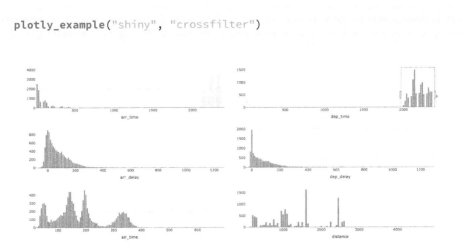

FIGURE 17.28: Cross-filtering six variables in the `flights` data from the **nycflights13** package (Wickham, 2018a). The filtering and binning logic occurs server-side resulting in a very minimal amount of data being sent between server and client (just the brush range and bar heights). Moreover, to perform the UI update, the client only has to tweak existing bar heights, so the overall user experience is quite responsive. For a video demonstration of the interactive, see `https://bit.ly/shiny-crossfilter`. For the interactive, see `https://plotly-r.com/interactives/shiny-crossfilter.html`

One weakness of a typical cross-filter interface like Figure 17.28 is that it's difficult to make visual comparisons. That is, when a filter is applied, you lose a visual reference to the overall distribution and/or prior filter, making it difficult to make meaningful comparisons. Figure 17.28 modifies the logic of Figure 17.29 to enable filter comparisons by adding the ability to change the color of the brush. Moreover, for the sake of demonstration and simplicity, it also allows for only one active filter per color (i.e., brushing within color is transient). One could borrow logic from Figure 17.28 to allow multiple filters for each color, but this would require multiple `brush_ranges`.

Since brushing within color is transient, in contrast to Figure 17.28, Figure 17.29 doesn't have to track the state of all the active brushes. It does, however, need to display *relative* rather than absolute frequencies to facilitate comparison of small filter range(s) to the overall distribution. This particular implementation takes the overall distribution as a "base layer" that remains fixed and overlays a handful of "selection layers", one for each possible brush color. These selection layers all begin with a height of 0. When a brush event occurs, only the bar height of the appropriate layer is updated. This approach may feel like a hack, but it leads to a fluid user experience because it's not much work to adjust the height of a bar that already exists.

```
plotly_example("shiny", "crossfilter_compare")
```

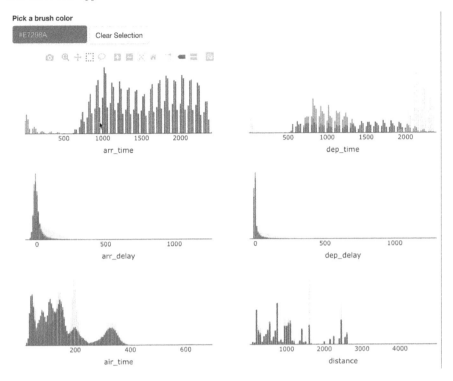

FIGURE 17.29: Comparing filters with a dynamic color brush. This particular example compares 'red eye' flights (in green) to early morning flights (in orange). This makes it easier to see that delays occur more often for 'red eye' flights. For a video demonstration of the interactive, see `https://bit.ly/shiny-crossfilter-compare`. For the interactive, see `https://plotly-r.com/interactives/shiny-crossfilter-compare.html`

17.4.3 A draggable brush

A *draggable* linked brush is great for conditioning via a moving window. For example, in a cross-filtering example like Figure 17.28, it would be nice to condition on a certain hour of day, then drag that hour interval along the axis to explore how the hourly distribution changes throughout the day. At the time of writing, plotly.js does not support a draggable brush, but we could implement one with a clever use of an editable rectangle shape. Figure 17.30 demonstrates this idea in a **shiny** application by drawing a rectangle shape that mimics the plotly.js

brush, then uses the `"plotly_relayout"` event to determine the limits
of the brush (instead of the `"plotly_brushed"` or `"plotly_brushing"`).

```
plotly_example("shiny", "drag_brush")
```

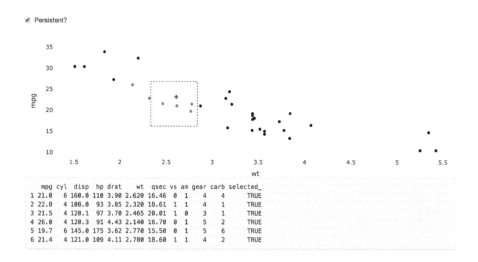

FIGURE 17.30: A draggable brush with both a transient and persistent
mode. The dragging ability is done by mimicking the native plotly.js
brush with an editable rectangle shape and listening to changes in
that brush. The implementation of transient vs. persistent mode is a
matter of forgetting or remembering previous state(s) of the brush.
For a video demonstration of the interactive, see https://bit.ly/shiny-
drag-brush. For the interactive, see https://plotly-r.com/interactives/
shiny-drag-brush.html

17.5 Discussion

Compared to the linking framework covered in Section 16.1, the
ability to link views server-side with **shiny** opens the door to
many new possibilities. This chapter focuses mostly on using just
plotly within **shiny**, but the **shiny** ecosystem is vast and these

techniques can of course be used to inform other views, such as static plots, other **htmlwidgets** packages (e.g., **leaflet**, **DT**, **network3D**, etc.), and other custom **shiny** bindings. In fact, I have numerous **shiny** apps publicly available via an R package that use numerous tools to provide exploratory interfaces to a variety of domain-specific problems, including `zikar::explore()` for exploring Zika virus reports, `eechidna::launch()` for exploring Australian election and census data, and `bcviz::launch()` for exploring housing and census information in British Colombia (Sievert, 2018a,d; Cook et al., 2017). These complex applications also serve as a reference as to how one can use the client-side linking framework (i.e., **crosstalk**) inside a larger **shiny** application. See this video for an example:

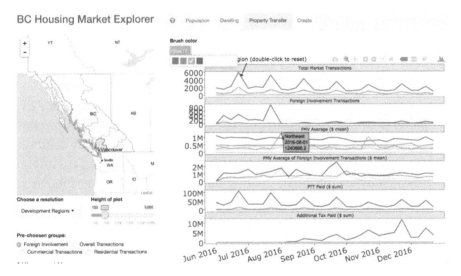

FIGURE 17.31: Example of a **shiny** app that has **crosstalk** functionality embedded. For a video demonstration of the interactive, see `https://bit.ly/shiny-crosstalk-examples`. For the interactive, see `https://plotly-r.com/interactives/shiny-crosstalk-examples.html`

Sometimes **shiny** gets a bad rap for being too slow or unresponsive, but as we learned in Sections 17.3.1 and 17.4, we can still have very advanced functionality as well as a good user experience; it just takes a bit more effort to optimize performance in some cases. In fact, one could argue that a server-client approach to cross-filtering, as done in Figure 17.28 is more scalable than a purely client-side approach since the client wouldn't need to know about the raw data, just the summary statistics. Nevertheless, sometimes linking views server-side simply isn't an option for you or your organization.

Maybe your IT administrator simply won't allow you to distribute your work outside of a standalone HTML file. Figure 17.11 is just one example of a linked graphic that *could* be replicated using the graphical querying framework from Section 16.1, but it would require pre-computing every possible view (which becomes un-manageable when there are many possible selections) and posing the update logic as a database query. When users are only allowed to select (e.g., click/hover) a single element at a time, the number of possible selections increases linearly with the number of elements, but when users are allowed to select any subset of elements (e.g., scatterplot brushing), the number of possible selections explodes (increases at a factorial rate). For example, adding a cell to Figure 17.11 only adds one possible selection, but if we added more states to Figure 17.11, the number of possible states goes from 50! to 51!.

Even in the case that you need a standalone HTML file and the R API that **plotly** provides doesn't support the type of interactivity that you desire, you can always layer on additional JavaScript to hopefully achieve the functionality you desire. This can be useful for something as simple as opening a hyperlink when clicking on marker of a plotly graph. This topic is covered in Chapter 18.

Part V

Event handling in JavaScript

18

Introduction

The same plotly.js events that we leveraged in **shiny** through `event_data()` in Section 17.2 can also be handled using JavaScript (JS) instead of R, which offers numerous advantages:

1. Your web browser natively understands JS, so writing event handlers in JS instead of R offers the potential of having a purely client-side webpage instead of client-server app, making the end result easier to share, host, and maintain.[1] That being said, the effort required to rewrite a 'heavy-duty' client-server app as a 'light-weight' client-side webpage isn't always worth the investment; so before doing so, you should have a clear vision for the interactivity you desire, be fairly confident that vision won't change in the future, and have a use-case that doesn't require sophisticated statistical computations to be run dynamically in response to user events (i.e., it's not practical or feasible to pre-compute).[2]
2. There are certain things you can do in JS that you can't necessarily do in R, such as accessing the web browser's `window` API to open hyperlinks in response to plotly.js click events (e.g., Figure 21.1).
3. JS event handlers can be noticeably faster than running comparable code on an external R process, especially with a slow internet connection.

---

[1]Comparatively speaking, client-server apps require way more runtime software dependencies. In the case of **shiny** apps, RStudio provides accessible resources for hosting shiny apps `https://shiny.rstudio.com/articles/#deployment`, but using these services to host apps that encounter lots of traffic will either cost money and/or time for setting up the proper computational infrastructure.

[2]Compared to JS, R has way more facilities for statistical computing.

For those new to JS, you may find it helpful to compare code examples from this part of the book to code examples from Section 17.3.1.[3] That's because, the `plotlyProxy()` interface which powers that section is just an R interface to plotly.js's JavaScript functions[4], and is primarily for updating graphs within an event handler. Therefore, if you understand how the examples in that section work, you can translate a good amount of that knowledge to a JS context as well. However, when handling these events in JS instead of R, you'll want to be familiar with JavaScript Object Notation (JSON), which is introduced in Chapter 19. That chapter also offers a minimal JS programming foundation for manipulating JSON, then Chapter 20 quickly covers how to attach JS event handlers to various plotly.js events, which is really all that's required to loosely understand the bulk of the examples in Chapters 21 and 22.

An important thing to know about when doing any sort of web development is how to open and navigate to the web browser's developer tools. Through the developer tools, you can access a JS console to run and test out JS code, inspect and debug the JS/CSS/HTML code behind a website, query components of the Document Object Model (DOM), inspect network traffic, and much more. In our use case of writing plotly.js event handlers, the JS console will come in handy especially to see what information a plotly.js event is firing (think of it as the analog of printing output to the R console in a shiny app), before writing the actual event handler. To open the console of a web browser (including RStudio), you can likely do: right-click -> "Inspect Element" -> "Console" tab (or similar).

One way to write a custom event handler for a **plotly** graph is to leverage the `onRender()` function from the **htmlwidgets** package. This R function accepts a JS function as a string and calls that function when the widget is done rendering in the browser. The JS function needs (at least) one argument, `el`, which is the DOM element containing the **plotly** graph. It's worth noting that `htmlwidgets::onRender()` serves a more general purpose than registering plotly.js event handlers, so you

---

[3]In fact, converting some examples from that section from **shiny** to JavaScript/HTML would be a good exercise!

[4]`https://plot.ly/javascript/plotlyjs-function-reference`

could use it to a bunch of other things, such as adding conditional logic based on information stored in el. Figure 18.1 shows how you could use onRender() to log (and inspect) the DOM element to your browser's JS console. To demonstrate some useful DOM element properties, Figure 18.1 uses Firefox to inspect the element as a global variable, but as Figure 20.1 shows, Chrome currently offers better tools for code debugging the JS function provided to onRender() (e.g., setting breakpoints in virtual memory).

```
library(htmlwidgets)
plot_ly(z = ~volcano) %>%
  onRender("function(el) { console.log(el); }")
```

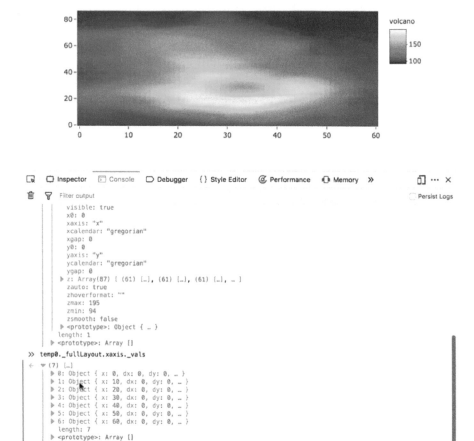

FIGURE 18.1: Using `htmlwidgets::onRender()` to inspect the relevant DOM instance containing the **plotly** graph and information related to its current display. The `_fullData` and `_fullLayout` attributes bound to the element are 'internal' (meaning relying on this information in production code is discouraged), but do provide a useful description of the chart's current state, especially if you need access to computations done by plotly.js (e.g., axis tick placement). For a video demonstration of the interactive, see `https://bit.ly/js-console-log`. For the interactive, see `https://plotly-r.com/interactives/console-log.html`

If you're completely new to JS and JSON, Chapter 19 provides a foundation for understanding the subsequent sections, but those who are already familiar can skip ahead to Chapter 20, which shows how to log plotly.js event data to the JS console via `htmlwidgets::onRender()`.

19

Working with JSON

JavaScript (JS) and other web technologies are intimidating and time-consuming to learn, but by borrowing some knowledge of R's data structures[1], we can get up and running with useful examples fairly quickly. JavaScript Object Notation (JSON) is a popular data-interchange format that JS uses to work with data. As it turns out, working with JSON in JS is somewhat similar to working with `list()`s in R; both are recursive and heterogeneous data structures that have similar semantics for accessing values. In JSON, there are three basic building blocks: objects, arrays, and primitive data types (e.g., number, string, Boolean, `null`, `undefined`).

Loosely speaking, a JSON array is similar to a un-named `list()` in R and a JSON object is similar to an un-named `list()`. In fact, if you're already comfortable creating and subsetting named and un-named `list()`s in R, you can transfer some of that knowledge to JSON arrays and objects.

19.1 Assignment, subsetting, and iteration

In R, the `<-` operator assigns a value to a name, and the `[[` operator extracts a list element by index:

```
arr <- list("hello", "world", 10)
arr[[1]]
#> "hello"
```

---

[1]If you'd like a nice succinct overview on the topic, see `http://adv-r.had.co.nz/Data-structures.html`

In JS, the = assigns a value to a name. When assigning a new name, you should include the var keyword (or similar) to avoid creation of a global variable. The [operator extracts list elements by index, but **be careful, indexing in JS starts at 0 (not 1)!**

```
var arr = ["hello", "world", 10];
arr[0]
// "hello"
```

In R, the $ and [[operator can be used to extract list elements by name. The difference is that $ does partial matching of names, while [[requires the exact name.

```
obj <- list(x = c("hello", "world"), zoo = 10)
obj$z
#> 10
obj[["zoo"]]
#> 10
```

In JS, the . and [operator can be used to extract list elements by name. In either case, the naming must be exact.

```
var obj = {
  x: ["hello", "world"],
  zoo: 10
}
obj.zoo
// 10
obj['zoo']
// 10
```

Unlike R list()s, arrays and objects in JS come with properties and methods that can be accessed via the . operator. Arrays, in particular, have a length property and a map() method for applying a function to each array element:

```
arr.length
// 3
arr.map(function(item) { return item + 1; });
// ["hello1", "world1", 11]
```

In R, both the `lapply()` and `purrr::map()` family of functions provide a similar functional interface. Also, note that operators like `+` in JS do even more type coercion than R, so although `item + 1` works for strings in JS, it would throw an error in R (and that's ok, most times you probably don't want to add a string to a number). If instead, you wanted to only add 1 to numeric values, you could use `is.numeric()` in R within an if else statement.

```
purrr::map(arr, function(item) {
  if (is.numeric(item)) item + 1 else item
})
#> [[1]]
#> [1] "hello"
#>
#> [[2]]
#> [1] "world"
#>
#> [[3]]
#> [1] 11
```

In JS, you can use the `typeof` keyword to get the data type as well as the conditional ternary operator (`condition ? exprT : exprF`) to achieve the same task.

```
arr.map(function(item) {
  return typeof item == "number" ? item + 1 : item;
});
// ["hello", "world", 11]
```

There are a handful of other useful array[2] and object[3] methods, but to keep things focused, we'll only cover what's required to comprehend in Chapter 20. A couple of examples in that section use the `filter()` method, which like `map()` applies a function to each array element, but expects a logical expression and returns only the elements that meet the condition.

```
arr.filter(function(item) { return typeof item == "string"; });
// ["hello", "world"]
```

19.2   Mapping R to JSON

In R, unlike JSON, there is no distinction between scalars and vectors of length 1. That means there is ambiguity as to what a vector of length 1 in R should map to in JSON. The **jsonlite** package defaults to an array of length 1, but this can be avoided by setting `auto_unbox = TRUE`.

```
jsonlite::toJSON("A string in R")
#> ["A string in R"]
jsonlite::toJSON("A string in R", auto_unbox = TRUE)
#> "A string in R"
```

It's worth noting that plotly.js, which consumes JSON objects, has specific expectations and rules about scalars versus arrays of length 1. If you're calling the plotly.js library directly in JS, as we'll see later in Chapter 20, you'll need to be mindful of the difference between scalars and arrays of length 1. Some attributes, like `text` and `marker.size`, accept both scalars and arrays and apply different rules based on the difference. Some other attributes, like `x`, `y`, and `z`, only accept arrays and will error out if given a scalar. To learn about these rules and expectations,

---

[2]https://developer.mozilla.org/en-US/docs/Web/JavaScript/Reference/Global_Objects/Array
[3]https://developer.mozilla.org/en-US/docs/Web/JavaScript/Reference/Global_Objects/Object

you can use the `schema()` function from R to inspect plotly.js's specification as shown in Figure 19.1. Note that attributes with a `val_type` of `'data_array'` require an array while attributes with an `arrayOk: true` field accept either scalars or arrays.

```
schema()
```

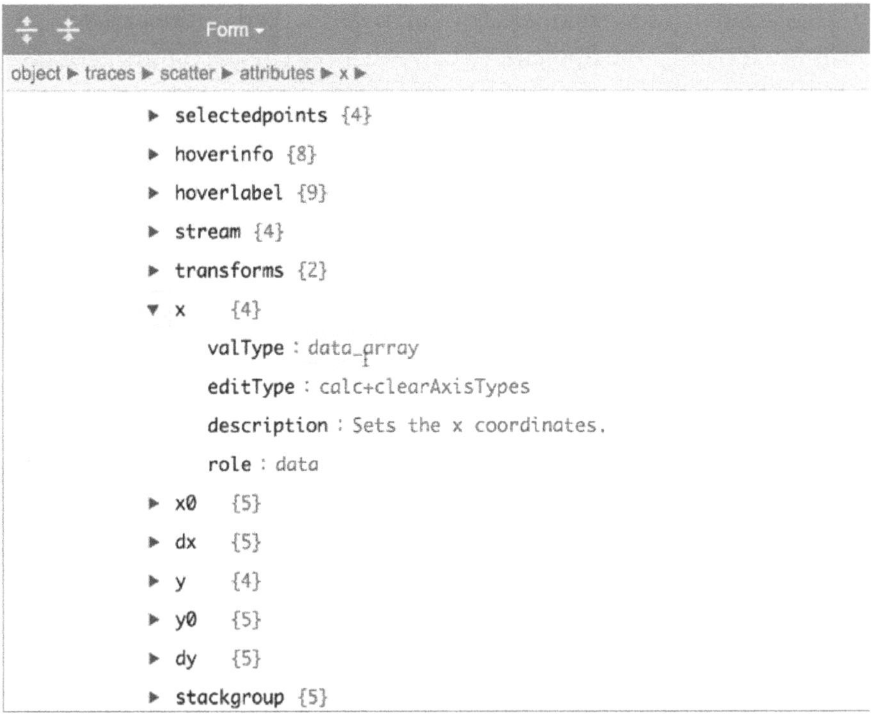

FIGURE 19.1: Using the plotly `schema()` to obtain more information about expected attribute types. For a video demonstration of the interactive, see `https://bit.ly/json_schema`. For the interactive, see `https://plotly-r.com/interactives/json-schema.html`

In JSON, unlike R, there is no distinction between a heterogeneous and homogeneous collection of data types. In other words, in R, there is an important difference between `list(1, 2, 3)` and `c(1, 2, 3)` (the latter is an atomic vector and has a different set of rules). In JSON, there is no strict notion of a homogeneous collection, so working with JSON arrays is essentially like being forced to use list() in R. This subtle fact can lead to some surprising results when trying to serialize R vectors as JSON arrays. For instance, if you wanted to create a JSON array, say [1,"a",true] using R objects, you may be tempted to do the following:

```
jsonlite::toJSON(c(1, "a", TRUE))
#> ["1","a","TRUE"]
```

But this actually creates an array of strings instead of the array with a number, string, and Boolean that we desire. The problems actually lie in the fact that c() coerces the collection of values into an atomic vector. Instead, you should use list() over c():

```
jsonlite::toJSON(list(1, "a", TRUE), auto_unbox = TRUE)
#> [1,"a",true]
```

20

Adding custom event handlers

When using onRender() to provide a JS function to be called upon static render of a **plotly** object, the relevant DOM element (el) has an on() method that accepts a function to be called whenever a plotly.js (or DOM) event occurs on that DOM element. Currently all plotly.js event handlers accept a function with a single argument, and that argument either contains nothing (e.g., "plotly_afterplot", etc.) or a single object with all the relevant information about the event (e.g., "plotly_hover", "plotly_selected", etc.). Figure 20.1 logs and inspects data (d) emitted during the "plotly_hover", "plotly_click", and "plotly_selected" events. The object emitted for these events includes a key, named points, with information tying the selection back to the input data. The points key is always an array of object(s) where each object represents a different data point. This object contains any supplied customdata, the relevant x/y location, and a reference back to the input data.

```
library(htmlwidgets)
plot_ly(mtcars, x = ~wt, y = ~mpg) %>%
  onRender("
    function(el) {
      el.on('plotly_hover', function(d) {
        console.log('Hover: ', d);
      });
      el.on('plotly_click', function(d) {
        console.log('Click: ', d);
      });
      el.on('plotly_selected', function(d) {
        console.log('Select: ', d);
      });
    }
  ")
```

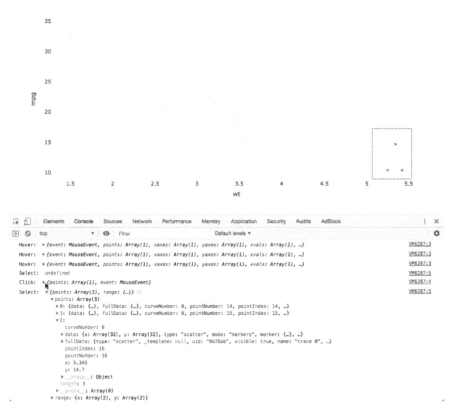

FIGURE 20.1: Inspecting event data for hover, click, and selected events. If a click or hover event does not derive from a statistical aggregation (e.g., boxplot, histogram, etc.), the `points` array is of length 1; otherwise, the length corresponds to how many input values are represented in the selection. In Chrome, when you log an object to the console, you can click on a link to the JS source function where you can then set breakpoints. For a video demonstration of the interactive, see `https://bit.ly/console-log-event`. For the interactive, see `https://plotly-r.com/interactives/console-log-event.html`

21

Supplying custom data

As covered in Section 17.2, it's often useful to supply meta-information (i.e., custom data) to graphical marker(s) and use that information when responding to a event. For example, suppose we'd like each point in a scatterplot to act like a hyperlink to a different webpage. In order to do so, we can supply a url to each point (as metadata) and instruct the browser to open the relevant hyperlink on a click event. Figure 21.1 does exactly this by supplying urls to each point in R through the customdata attribute and defining a custom JS event to window.open() the relevant url upon a click event. In this case, since each point represents one row of data, the d.point is an array of length 1, so we may obtain the url of the clicked point with d.points[0].customdata.

```r
library(htmlwidgets)

p <- plot_ly(mtcars, x = ~wt, y = ~mpg) %>%
  add_markers(
    text = rownames(mtcars),
    customdata = paste0("http://google.com/#q=", rownames(mtcars))
  )

onRender(
  p, "
  function(el) {
    el.on('plotly_click', function(d) {
      var url = d.points[0].customdata;
      window.open(url);
    });
  }
")
```

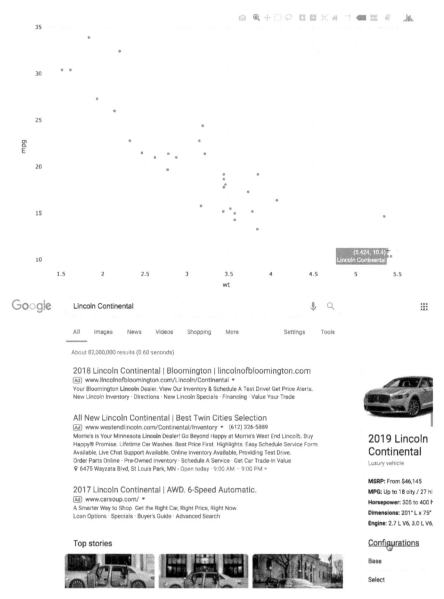

FIGURE 21.1: Attaching hyperlinks to each point in a scatterplot and using a custom JS event to open that Google search query upon clicking a point. For a video demonstration of the interactive, see https://bit.ly/click-open. For the interactive, see https://plotly-r.com/interactives/click-open.html

In addition to using `window.open()` to open the url, we could also add it to the plot as an annotation using the plotly.js function `Plotly.relayout()`, as done in Figure 21.2. Moreover, since plotly annotations support HTML markup, we can also treat that url as a true HTML hyperlink by wrapping it in an HTML `<a>` tag. In cases where your JS function starts to get complex, it can help to put that JS function in its own file, then use the R function `readLines()` to read it in as a string and pass along `onRender()` as done below:

```
onRender(p, readLines("js/hover-hyperlink.js"))
```

```js
// Start of the hover-hyperlink.js file
function(el) {
  el.on('plotly_hover', function(d) {
    var url = d.points[0].customdata;
    var ann = {
      text: "<a href='" + url + "'>" + url + "</a>",
      x: 0,
      y: 0,
      xref: "paper",
      yref: "paper",
      yshift: -40,
      showarrow: false
    };
    Plotly.relayout(el.id, {annotations: [ann]});
  });
}
```

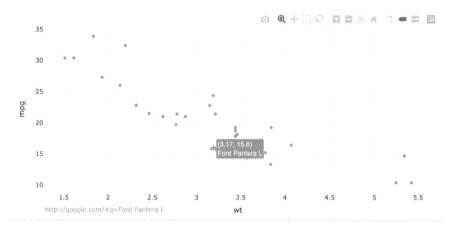

FIGURE 21.2: Using `Plotly.relayout()` to add and change hyperlink in response to hover events. For a video demonstration of the interactive, see `https://bit.ly/hover-annotate`. For the interactive, see `https://plotly-r.com/interactives/hover-annotate.html`

 When using `Plotly.relayout()`, or any other plotly.js function to modify a plot, you'll need to know the id attribute of the relevant DOM instance that you want to manipulate. When working with a single object, you can simply use `el.id` to access the id attribute of that DOM instance. However, when trying to target another object, it gets trickier because id attributes are randomly generated by **htmlwidgets**. In that case, you likely want to pre-specify the id attribute so you can reference it client-side. You can pre-specify the id for any **htmlwidgets** object, say `widget`, by doing `widget$elementId <- "myID"`.

The `customdata` attribute can hold any R object that can be serialized as JSON, so you could, for example, attach complex data to markers/lines/text/etc. using base64 strings. This could be useful for a number of things such as displaying an image on hover or click. For security reasons, plotly.js doesn't allow inserting images in the tooltip, but you can always define your own tooltip by hiding the tooltip (`hoverinfo='none'`), then populating your own tooltip with suitable manipulation of the DOM in response to `"plotly_hover"`/`"plotly_unhover"` events. Figure 21.3 demonstrates how to leverage this infrastructure to display a PNG image in the top-left corner of a graph whenever a text label is hovered upon.[1]

```r
x <- 1:3
y <- 1:3
logos <- c("r-logo", "penguin", "rstudio")
# base64 encoded string of each image
uris <- purrr::map_chr(
  logos, ~ base64enc::dataURI(file = sprintf("images/%s.png", .x))
)
# hoverinfo = "none" will hide the plotly.js tooltip, but the
# plotly_hover event will still fire
plot_ly(hoverinfo = "none") %>%
  add_text(x = x, y = y, customdata = uris, text = logos) %>%
  htmlwidgets::onRender(readLines("js/tooltip-image.js"))
```

```js
// Start of the tooltip-image.js file
// inspired, in part, by https://stackoverflow.com/a/48174836/1583084
function(el) {
  var tooltip = Plotly.d3.select('#' + el.id + ' .svg-container')
    .append("div")
    .attr("class", "my-custom-tooltip");

  el.on('plotly_hover', function(d) {
    var pt = d.points[0];
    // Choose a location (on the data scale) to place the image
    // Here I'm picking the top-left corner of the graph
    var x = pt.xaxis.range[0];
    var y = pt.yaxis.range[1];
    // Transform the data scale to the pixel scale
    var xPixel = pt.xaxis.l2p(x) + pt.xaxis._offset;
    var yPixel = pt.yaxis.l2p(y) + pt.yaxis._offset;
    // Insert the base64 encoded image
```

---

[1]As long as you are not allowing down-stream users to input paths to the input files (e.g., in a **shiny** app), you shouldn't need to worry about the security of this example.

```
    var img = "<img src='" +  pt.customdata + "' width=100>";
    tooltip.html(img)
      .style("position", "absolute")
      .style("left", xPixel + "px")
      .style("top", yPixel + "px");
    // Fade in the image
    tooltip.transition()
      .duration(300)
      .style("opacity", 1);
  });

  el.on('plotly_unhover', function(d) {
    // Fade out the image
    tooltip.transition()
      .duration(500)
      .style("opacity", 0);
  });
}
```

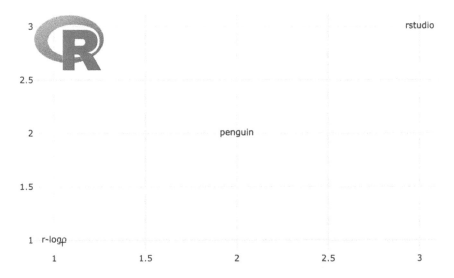

FIGURE 21.3: Displaying an image on hover in a scatterplot. For a video demonstration of the interactive, see `https://bit.ly/tooltip-image`. For the interactive, see `https://plotly-r.com/interactives/tooltip-image.html`

It's worth noting that the JavaScript that powers Figure 21.3 works for other Cartesian charts, even `heatmap` (as shown in Figure 21.4), but it would need to be adapted for 3D chart types.

```r
plot_ly(hoverinfo = "none") %>%
  add_heatmap(
    z = matrix(1:9, nrow = 3),
    customdata = matrix(uris, nrow = 3, ncol = 3)
  ) %>%
  htmlwidgets::onRender(readLines("js/tooltip-image.js"))
```

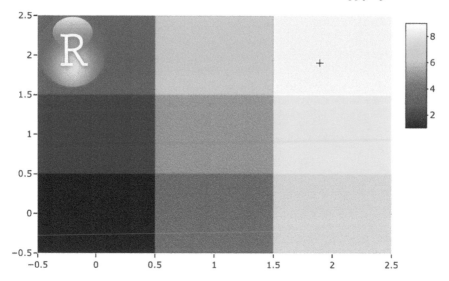

FIGURE 21.4: Displaying an image on hover in a heatmap. For a video demonstration of the interactive, see `https://bit.ly/tooltip-image-heatmap`. For the interactive, see `https://plotly-r.com/interactives/tooltip-image-heatmap.html`

On the JS side, the `customdata` attribute is designed to support *any* JS array of appropriate length, so if you need to supply numerous custom values to particular marker(s), list-columns in R provide a nice way to do so. Figure 21.5 leverages this idea to bind both the `city` and `sales` values to each point along a time series and display those values on hover. It also demonstrates how one can use the graphical querying framework from Section 16.1 in tandem with a custom JS event. That is, `highlight_key()` and `highlight()` control the highlighting of the time series, while the custom JS event adds the plot annotation (all based on the same `"plotly_hover"` event). In this case, the highlighting, annotations, and circle shapes are triggered by a `"plotly_hover"` event and they all work in tandem because event handlers are cumulative. That means, if you wanted, you could register multiple custom handlers for a particular event.

```r
library(purrr)

sales_hover <- txhousing %>%
  group_by(city) %>%
  highlight_key(~city) %>%
  plot_ly(x = ~date, y = ~median, hoverinfo = "name") %>%
  add_lines(customdata = ~map2(city, sales, ~list(.x, .y))) %>%
  highlight("plotly_hover")

onRender(sales_hover, readLines("js/tx-annotate.js"))
```

```js
// Start of the tx-annotate.js file
function(el) {
  el.on("plotly_hover", function(d) {
    var pt = d.points[0];
    var cd = pt.customdata;
    var num = cd[1] ? cd[1] : "No";
    var ann = {
      text: num + " homes were sold in "+cd[0]+", TX in this month",
      x: 0.5,
      y: 1,
      xref: "paper",
      yref: "paper",
      xanchor: "middle",
      showarrow: false
    };
    var circle = {
      type: "circle",
      xanchor: pt.x,
      yanchor: pt.y,
      x0: -6,
      x1: 6,
      y0: -6,
      y1: 6,
      xsizemode: "pixel",
      ysizemode: "pixel"
```

```
    };
    Plotly.relayout(el.id, {annotations: [ann], shapes: [circle]});
  });
}
```

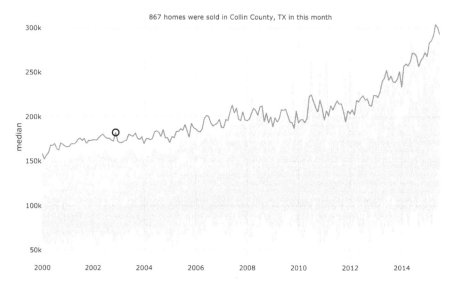

FIGURE 21.5: Combining the graphical querying framework from Section 16.1 with a custom JS event handler to highlight a time series as well as circling the month selected. This example supplies a list-column to customdata in order to populate an informative title based on the user's selection of city and month. For a video demonstration of the interactive, see https://bit.ly/tx-annotate. For the interactive, see https://plotly-r.com/interactives/tx-annotate.html

Sometimes supplying and accessing customdata alone is not quite enough for the task at hand. For instance, what if we wish to add the average monthly sales to the annotation for the city of interest in Figure 21.5? In cases like this, we may need to use customdata to query a portion of the plot's input data, like Figure 21.5 does to compute and display average sales for a given city. This implementation leverages the fact that each selected point (pt) contains a reference to the entire trace it derives from (pt.data). As discussion behind Figure 3.2 noted, this particular plot has a *single trace* and uses missing values to create separate lines for each city. As a result, pt.data.customdata contains all the customdata we supplied from the R side, so to get all the sales for a given city, we first need to filter that array down to only the elements that belong to that city (while being careful of missing values!).

```
onRender(sales_hover, readLines("js/tx-mean-sales.js"))
```

```js
// Start of the tx-mean-sales.js file
function(el) {
  el.on("plotly_hover", function(d) {
    var pt = d.points[0];
    var city = pt.customdata[0];

    // get the sales for the clicked city
    var cityInfo = pt.data.customdata.filter(function(cd) {
      return cd ? cd[0] == city : false;
    });
    var sales = cityInfo.map(function(cd) { return cd[1] });

    // yes, plotly bundles d3 which you can access via Plotly.d3
    var avgsales = Math.round(Plotly.d3.mean(sales));

    // Display the mean sales for the clicked city
    var ann = {
      text: "Mean monthly sales for " + city + " is " + avgsales,
      x: 0.5,
      y: 1,
      xref: "paper",
      yref: "paper",
      xanchor: "middle",
      showarrow: false
    };
    Plotly.relayout(el.id, {annotations: [ann]});
```

```
    });
}
```

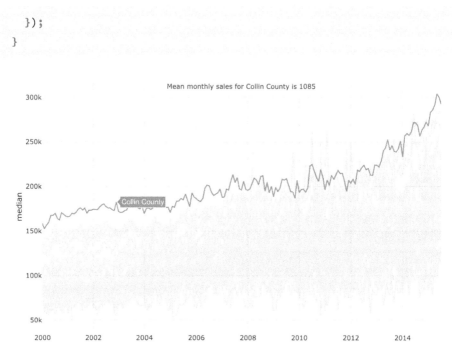

FIGURE 21.6: Displaying the average monthly sales for a city of interest on hover. This implementation supplies all the raw sales figures, then uses the hovered `customdata` value to query sales for the given city and display the average. For a video demonstration of the interactive, see `https://bit.ly/tx-mean-sales`. For the interactive, see `https://plotly-r.com/interactives/tx-mean-sales.html`

Figure 21.7 uses the same `customdata` supplied to Figure 21.6 in order to display a histogram of monthly sales for the relevant city on hover. In addition, it displays a vertical line on the histogram to reflect the monthly sales for the point closest to the mouse cursor. To do all this efficiently, it's best to add the histogram trace on the first hover event using `Plotly.addTraces()`, then supply different `sales` data via `Plotly.restyle()` (generally speaking, `restyle()` is way less expensive than `addTraces()`). That's why the implementation leverages the fact that the DOM element (`el`) contains a reference to the current graph data (`el.data`). If the current graph has a trace with a type of histogram,

then it adds a histogram trace; otherwise, it supplies new x values to
the histogram.

```
sales_hover %>%
  onRender(readLines("js/tx-annotate.js")) %>%
  onRender(readLines("js/tx-inset-plot.js"))



// Start of the tx-inset-plot.js file
function(el) {
  el.on("plotly_hover", function(d) {
    var pt = d.points[0];
    var city = pt.customdata[0];

    // get the sales for the clicked city
    var cityInfo = pt.data.customdata.filter(function(cd) {
      return cd ? cd[0] == city : false;
    });
    var sales = cityInfo.map(function(cd) { return cd[1] });

    // Collect all the trace types in this plot
    var types = el.data.map(function(trace) { return trace.type; });
    // Find the array index of the histogram trace
    var histogramIndex = types.indexOf("histogram");

    // If the histogram trace already exists, supply new x values
    if (histogramIndex > -1) {

      Plotly.restyle(el.id, "x", [sales], histogramIndex);

    } else {

      // create the histogram
      var trace = {
        x: sales,
        type: "histogram",
        marker: {color: "#1f77b4"},
        xaxis: "x2",
```

```
      yaxis: "y2"
    };
    Plotly.addTraces(el.id, trace);

    // place it on "inset" axes
    var x = {
      domain: [0.05, 0.4],
      anchor: "y2"
    };
    var y = {
      domain: [0.6, 0.9],
      anchor: "x2"
    };
    Plotly.relayout(el.id, {xaxis2: x, yaxis2: y});


  }


  // Add a title for the histogram
  var ann = {
    text: "Monthly house sales in " + city + ", TX",
    x: 2003,
    y: 300000,
    xanchor: "middle",
    showarrow: false
  };
  Plotly.relayout(el.id, {annotations: [ann]});


  // Add a vertical line reflecting sales for the hovered point
  var line = {
    type: "line",
    x0: pt.customdata[1],
    x1: pt.customdata[1],
    y0: 0.6,
    y1: 0.9,
    xref: "x2",
    yref: "paper",
    line: {color: "black"}
```

```
    };
    Plotly.relayout(el.id, {'shapes[1]': line});
  });
}
```

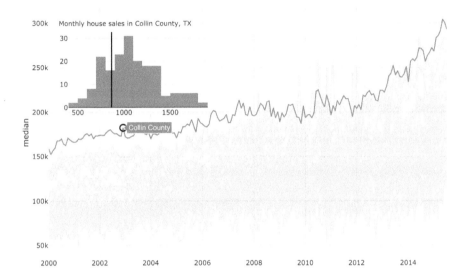

FIGURE 21.7: Adding another event handler to Figure 21.5 to draw an inset plot showing the distribution of monthly house sales. For a video demonstration of the interactive, see `https://bit.ly/tx-inset-plot`. For the interactive, see `https://plotly-r.com/interactives/tx-inset-plot.html`

22

Leveraging web technologies from R

22.1 Web infrastructure

Sometimes supplying `customdata` isn't the best way to achieve a particular interactive feature. In those cases, you likely want to leverage other R lower-level interfaces to web technologies. Recall from Section 13.2 that **htmlwidgets** objects are a special case of **htmltools** tags. That means you can always complement your widget(s) with arbitrary HTML content by adding additional tags. Figure 22.1 leverages this idea to place an empty HTML `<div>` container below the correlation heatmap which is then populated with a **plotly** scatterplot upon clicking a cell. As it turns out, you *could* implement Figure 22.1 by binding x/y data to each heatmap cell via `customdata`, but that would require the browser to store twice the amount of data as what's required here. Instead, this approach serializes the input data (`mtcars`) into a JSON file via **jsonlite** so the webpage can read and parse the full dataset once and select just the two required columns when required (on click). There are a lot of ways to read JSON in JavaScript, but here we use the d3.js library's `d3.json()` since **plotly** already comes bundled with the library (Bostock et al., 2011). Also, since the HTML file is reading the JSON from disk, most browsers won't render the HTML file directly (at least, by default, for security reasons). To get around that, we can start up a simple web server from R using **servr** to serve both the HTML and JSON in a way that your browser will deem safe to run (Xie, 2016).

```
library(plotly)
library(htmltools)

nms <- names(mtcars)
```

```r
p <- plot_ly(colors = "RdBu") %>%
  add_heatmap(
    x = nms,
    y = nms,
    z = ~round(cor(mtcars), 3)
  ) %>%
  onRender("
    function(el) {
      Plotly.d3.json('mtcars.json', function(mtcars) {
        el.on('plotly_click', function(d) {
          var x = d.points[0].x;
          var y = d.points[0].y;
          var trace = {
            x: mtcars[x],
            y: mtcars[y],
            mode: 'markers'
          };
          Plotly.newPlot('filtered-plot', [trace]);
        });
      });
    }
")



# In a temporary directory, save the mtcars dataset as json and
# the html to an index.html file, then open via a web server
withr::with_path(tempdir(), {
  jsonlite::write_json(as.list(mtcars), "mtcars.json")
  html <- tagList(p, tags$div(id = 'filtered-plot'))
  save_html(html, "index.html")
  if (interactive()) servr::httd()
})
```

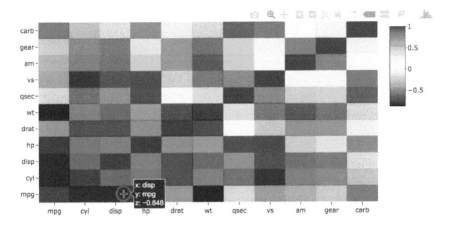

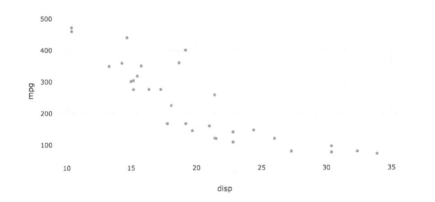

FIGURE 22.1: Clicking on a correlation matrix to populate a scatter-plot. For a video demonstration of the interactive, see `https://bit.ly/correlation-client-side`. For the interactive, see `https://plotly-r.com/interactives/correlation-client-side.html`

22.2 Modern JS and React

All the JavaScript (JS) we've seen thus far is natively supported by modern web browsers, but for larger projects, you may want to leverage modern versions of JS (i.e., ES6, ES7, etc.) and modern JS development tools (e.g., Babel, Webpack, etc.) for compiling modern JS to a version that all browsers can support (i.e., ES2015). The current landscape of JS development tooling is large, complex, fragmented, difficult for non-experts to navigate, and mostly beyond the scope of this book. However, thanks to R packages like **V8**, **reactR**, and **runpkg**, it turns out we can effectively leverage React[1] components[2] from R without fussing with system commands or setting up a complicated JS build toolchain.

The R package **runpkg** makes it easy to download any npm[3] (the main repository network for JS) package (via `https://unpkg.com/`) and include it in a webpage generated through the **htmltools** package (Sievert, 2019b). It does so by returning a `htmltools::htmlDependency()` object which encapsulates the downloaded files and includes the JS scripts (or CSS stylesheets) into any page that depends on that object. Here we use it to download a standalone bundle of a React library for rendering all sorts of different video formats, called `react-player`.

```
react_player <- runpkg::download_files(
  "react-player",
  "dist/ReactPlayer.standalone.js"
)
```

This `react-player` library provides a function called `renderReactPlayer()` that requires a placeholder (i.e., a DOM element) for inserting the video as well as a url (or file path) to the video. Figure 22.2 demonstrates how we could use it to render a YouTube video in response to a **plotly** click event:

```
library(htmltools)

# the video placeholder
video <- tags$div(id = "video", align = "center")

# upon clicking the marker, populate a video
```

---

[1]React is a modern JavaScript library, backed by Facebook, for building and distributing components of a website – `https://reactjs.org/`
[2]There are thousands of React components available. To get a sense of what's available, see this list `https://github.com/brillout/awesome-react-components`.
[3]`https://www.npmjs.com/`

```r
# in the DOM element with an id of 'video'
p <- plot_ly(x = 1, y = 1, size = I(50)) %>%
  add_text(
    text = emo::ji("rofl"),
    customdata = "https://www.youtube.com/watch?v=oHg5SJYRHA0",
    hovertext = "Click me!",
    hoverinfo = "text"
  ) %>%
  onRender(
    "function(el) {
       var container = document.getElementById('video');
       el.on('plotly_click', function(d) {
         var url = d.points[0].customdata;
         renderReactPlayer(container, {url: url, playing: true});
       })
     }"
  )

# create the HTML page
browsable(tagList(p, video, react_player))
```

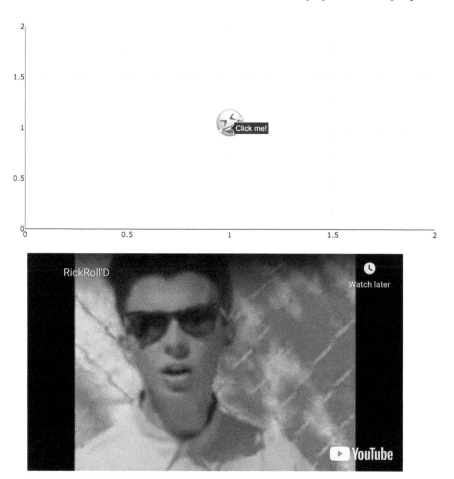

FIGURE 22.2: Using the `react-player` React component to render a video in response to a `'plotly_click'` event. For a video demonstration of the interactive, see https://bit.ly/react-player. For the interactive, see https://plotly-r.com/interactives/react-player.html

This `react-player` React library is rather unique in that it provides a standalone function, `renderReactPlayer()`, that enables rendering of a React component without loading React itself or leveraging special React syntax like JSX. It's more likely that the React component library will explicitly require you to import both React and React-DOM. You could use **runpkg** to download these React/ReactDOM as well, but the `html_dependency_react()` function from **reactR** package makes this even easier (Inc et al., 2019). Furthermore, **reactR** provides a `babel_transform()` function which will compile modern JS (e.g., ES6, ES2017, etc.) as well as special React markup (e.g., JSX) to a version of JS that all browsers support (e.g., ES5). For a toy example, Figure 22.3 demonstrates how one could leverage ES6, React, and React's JSX syntax to populate a `<h1>` title filled with a `customdata` message in response to a **plotly** click event.

```r
library(reactR)

# a placeholder for our react 'app'
app <- tags$div(id = "app")

p <- plot_ly(x = 1, y = 1) %>%
  add_markers(customdata = "Powered by React") %>%
  onRender(babel_transform(
    "el => {
      el.on('plotly_click', d => {
        let msg = d.points[0].customdata;
        ReactDOM.render(
          <h1>{msg}</h1>,
          document.getElementById('app')
        )
      })
    }"
  ))

# create the HTML page
browsable(tagList(p, app, html_dependency_react()))
```

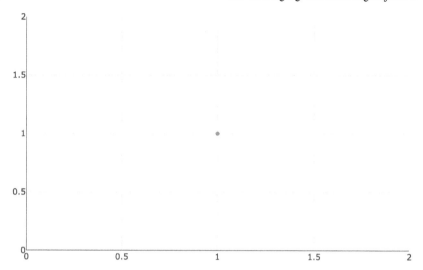

Powered by React

FIGURE 22.3: Using `babel_transform()` to leverage ES6, React, and JSX. For a video demonstration of the interactive, see `https://bit.ly/babel-transform`. For the interactive, see `https://plotly-r.com/interactives/babel.html`

For a more serious example, we could leverage another React component, named `react-data-grid`, to display the data within a **plotly** scatterplot brush, as done in Figure 22.4. Again, we can use **runpkg** to download a bundle of `react-data-grid`, but this library doesn't come with `React`/`ReactDOM`, so we must explicitly include it this time around. In fact, this approach of explicitly importing and calling `ReactDOM.render()` on your component is a more common approach than the custom standalone interface approach (i.e., `renderReactPlayer()`) used in Figure 22.2.

```
data_grid_js <- runpkg::download_files(
  "react-data-grid",
  "dist/react-data-grid.min.js"
)
```

```r
# the data table placeholder
data_grid <- tags$div(id = "data-grid")

# upon clicking the marker, populate a video
# in the DOM element with an id of 'video'
p <- plot_ly(mtcars, x = ~wt, y = ~mpg) %>%
  add_markers(customdata = row.names(mtcars)) %>%
  layout(dragmode = "select") %>%
  onRender(babel_transform(
    "el => {
      var container = document.getElementById('data-grid');
      var columns = [
        {key: 'x', name: 'Weight'},
        {key: 'y', name: 'MPG'},
        {key: 'customdata', name: 'Model'}
      ];
      el.on('plotly_selecting', d => {
        if (d.points) {
          var grid = <ReactDataGrid
            columns={columns}
            rowGetter={i => d.points[i]}
            rowsCount={d.points.length}
          />;
          ReactDOM.render(grid, container);
        }
      });
      el.on('plotly_deselect', d => {
        ReactDOM.render(null, container);
      });
    }"
  ))

# create the HTML page
browsable(
  tagList(p, data_grid, html_dependency_react(), data_grid_js)
)
```

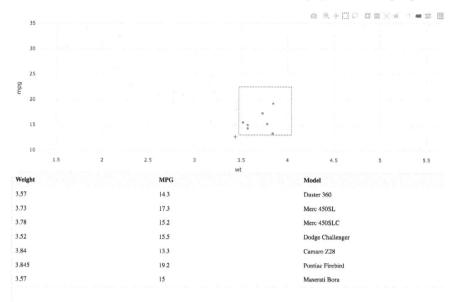

Weight	MPG	Model
3.57	14.3	Duster 360
3.73	17.3	Merc 450SL
3.78	15.2	Merc 450SLC
3.52	15.5	Dodge Challenger
3.84	13.3	Camaro Z28
3.845	19.2	Pontiac Firebird
3.57	15	Maserati Bora

FIGURE 22.4: Using the `react-data-grid` React component to render a data table of the observation within a scatterplot brush. For a video demonstration of the interactive, see https://bit.ly/react-data-grid. For the interactive, see https://plotly-r.com/interactives/react-data-grid.html

Part VI

Various special topics

23

Is plotly free and secure?

Yes! Both the R package and the underlying JavaScript library (plotly.js) are MIT licensed. That means you are free to distribute and commercialize anything you create with **plotly**. Optionally, you can sign up for a `https://plot.ly` account and use the `api_create()` function to upload **plotly** graphs to your account, but *a `https://plot.ly` account is not required to use **plotly***.

Moreover, the data behind a **plotly** graph won't leave the computer you're viewing it on, unless:

- You sign up for a plotly cloud[1] account and use the `api_create()` function to upload your graph or data. This service is free for public-facing graphs, but private hosting costs money.
- You add the 'send data to cloud' modebar button. By default, this button is not included, but you can add it by doing `config(p, cloud = TRUE)`.

In other words, unless you specifically request otherwise, your graph runs entirely offline (i.e., no requests to external services are made). That means you can be confident that your data is not being shared with anyone/anything without your knowledge.

---

[1]`https://plot.ly/products/cloud/`

24

Improving performance

Recall, from Figure 2.5, when you print a **plotly** object (or really any plot), there are two classes of performance to be aware of: print-time (i.e., build) and runtime (i.e., render). That is, build time can be classified as the time it takes for the object to be serialized as JSON/HTML, whereas run time is the time it takes for the browser to render the HTML into a webpage. In the case of **plotly**, there are two quick and easy things you can do to improve run time performance in *any context*:

- `toWebGL()`: This function attempts to render the chart using WebGL (i.e., Canvas) instead of Scalable Vector Graphics (SVG). The difference between these contexts is somewhat analogous to the difference between saving a static chart to PNG/JPG (pixel based) versus PDF (vector based). Vector-based graphics have the desirable property of producing sharp visuals that scale well to any size, but they don't scale well in the number of vectors (e.g., points, lines, polygons, etc.) that they need to render.

- `partial_bundle()`: This function attempts to reduce the size of the plotly.js bundle used to render the **plotly** graphs. The size of the default (i.e., main) plotly.js bundle is about 3MB, which can take a considerable amount of time to download with a slow internet connection, potentially leading to noticeable lag in initial page load for consumers of the graph. As it turns out, the main bundle is not always necessary to render every graph on a given website, so plotly.js provides partial bundles[1] that can render certain subsets of the graphing library. For instance, if you only need scatter, bar, or pie trace types, you can get away with the basic bundle which is currently under 1MB in size. This function is always safe to use when rendering a single **plotly** graph

---

[1] https://github.com/plotly/plotly.js/tree/master/dist#partial-bundles

in a webpage, but when rendering multiple graphs, you should take care not to include multiple bundles in the same page.

These two options may improve runtime performance without much of any thinking, but sometimes it's worth being more thoughtful about your visualization strategy by leveraging summaries (e.g., Section 13.3, Figure 17.27, and Figure 22.1) as well as being more explicit about how a graph responds to changes in the underlying data (e.g., Section 17.3.1). Mastering these more broad and complex subjects is critical for scaling interactive visualizations to truly large data[2], especially in the case of linking multiple views, where computational 'tricks' such as pre-aggregating distributive (e.g., min, max, sum, count) and algebraic (e.g., mean, var, etc.) statistics intelligently is a trademark of systems that enable real-time graphical queries of massive datasets (Liu et al., 2013; Lins et al., 2013; Moritz et al., 2019). As Wickham (2013) points out, it's also important to consider the uncertainty in these computationally efficient statistics, as they aren't nearly as statistically robust as their holistic counterparts (e.g., mean vs. median) that are more computationally intensive.

Since latency in interactive graphics is known to make exploratory data analysis a more challenging task (Liu and Heer, 2014), systems that optimize run over build performance are typically preferable. This is especially true for visualizations that *others* are consuming, but in a typical EDA context, where the person creating the visualization is the main consumer, build time performance is also an important factor because it also presents a hurdle to the analytical thought process. It's hard to give general advice on improving build-time performance in general, but a great first step in doing so is to profile the speed of your R code with something like the **profvis** package. This will at least let you know if the slowness you're experiencing is due to your own R code.

---

[2]Large data means different things to different people at different time periods. At the time of writing, I'd consider hundreds of millions of observations with at least a handful of variables to be large data.

25

Controlling tooltips

25.1 `plot_ly()` tooltips

There are two main approaches to controlling the tooltip: `hoverinfo` and `hovertemplate`. I suggest starting with the former approach since it's simpler, more mature, and enjoys universal support across trace types. On the other hand, `hovertemplate` does offer a convenient approach for flexible control over tooltip text, so it can be useful as well.

The `hoverinfo` attribute controls what other plot attributes are shown into the tooltip text. The default value of `hoverinfo` is `x+y+text+name` (you can verify this with `schema()`), meaning that plotly.js will use the relevant values of `x`, `y`, `text`, and `name` to populate the tooltip text. As Figure 25.1 shows, you can supply custom text (without the other 'calculated values') by supplying a character string `text` and setting `hoverinfo = "text"`. The character string can include Glyphs, unicode characters, and some (white-listed) HTML entities and tags.[1] At least currently, plotly.js doesn't support rendering of LaTeX[2] or images in the tooltip, but as demonstrated in Figure 21.3, if you know some HTML/JavaScript, you can always build your own custom tooltip.

```
library(tibble)
library(forcats)

tooltip_data <- tibble(
  x = " ",
  y = 1,
```

---

[1]If you find a tag or entity that you want that isn't supported, please request it to be added in the plotly.js repository https://github.com/plotly/plotly.js/issues/new
[2]https://github.com/plotly/plotly.js/issues/559

```r
  categories = as_factor(c(
    "Glyphs", "HTML tags", "Unicode",
    "HTML entities", "A combination"
  )),
  text = c(
    " glyphs _",
    "Hello <span style='color:red'><sup>1</sup><sub>2</sub></span>
    fraction",
    "\U0001f44b unicode \U00AE \U00B6 \U00BF",
    "&mu; &plusmn; &amp; &lt; &gt;   &times; &plusmn; &deg;",
    paste("<b>Wow</b> <i>much</i> options", emo::ji("dog2"))
  )
)


plot_ly(tooltip_data, hoverinfo = "text") %>%
  add_bars(
    x = ~x,
    y = ~y,
    color = ~fct_rev(categories),
    text = ~text
  ) %>%
  layout(
    barmode ="stack",
    hovermode = "x"
  )
```

FIGURE 25.1: Customizing the tooltip by supplying glyphs, Unicode, HTML markup to the text attributes and restricting displayed attributes with hoverinfo='text'.

Whenever a fill is relevant (e.g., add_sf(), add_polygons(), add_ribbons(), etc.), you have the option of using the hoveron attribute to generate a tooltip for the supplied data points, the filled polygon that those points define, or both. As Figure 25.2 demonstrates, if you want a tooltip attached to a fill, you probably want text to be of length 1 for a given trace. On the other hand, if you want to each point along a fill to have a tooltip, you probably want text to have numerous strings.

```r
p <- plot_ly(
  x = c(1, 2, 3),
  y = c(1, 2, 1),
  fill = "toself",
  mode = "markers+lines",
  hoverinfo = "text"
)

subplot(
  add_trace(p, text = "triangle", hoveron = "fills"),
```

```
  add_trace(p, text = paste0("point", 1:3), hoveron = "points")
)
```

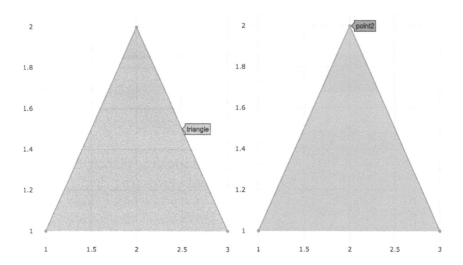

FIGURE 25.2: Using the `hoveron` attribute to control whether a tooltip is attached to fill or each point along that fill.

You can't supply custom text in this way to a statistical aggregation, but there are ways to control the formatting of values computed and displayed by plotly.js (e.g., x, y, and z). If the value that you'd like to format corresponds to an axis, you can use `*axis.hoverformat`. The syntax behind `hoverformat` follows d3js' format conventions. For numbers, see: https://github.com/d3/d3-format/blob/master/README.md#locale_format and for dates see: https://github.com/d3/d3-time-format/blob/master/README.md#locale_format

```
set.seed(1000)
plot_ly(x = rnorm(100), name = " ") %>%
  add_boxplot(hoverinfo = "x") %>%
  layout(xaxis = list(hoverformat = ".2f"))
```

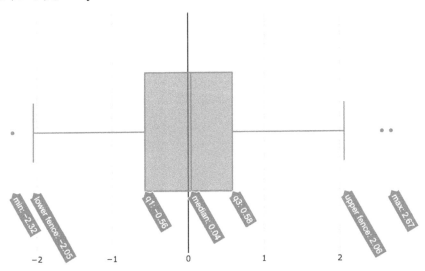

FIGURE 25.3: Using `xaxis.hoverformat` to round aggregated values displayed in the tooltip to two decimal places.

Computed values that don't have a corresponding axis likely have a `*hoverformat` trace attribute. Probably the most common example is the `z` attribute in a `heatmap` or `histogram2d` chart. Figure 25.4 shows how to format `z` values to have one decimal.

```
plot_ly(z = ~volcano) %>%
  add_heatmap(zhoverformat = ".1f") %>%
  layout(xaxis = list(hoverformat = ".2f"))
```

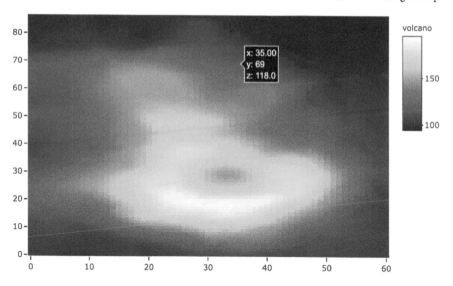

FIGURE 25.4: Formatting the displayed z values in a heatmap using zhoverformat.

It's admittedly difficult to remember where to specify these hoverformat attributes, so if you want a combination of custom text and formatting of computed values you can use hovertemplate, which overrides hoverinfo and allows you to fully specify the tooltip in a single attribute. It does this through special markup rules for inserting and formatting data values inside a string. Figure 25.5 provides an example of inserting x and y in the tooltip through the special %{variable:format} markup as well as customization of the secondary box through <extra> tag. For a full description of this attribute, including the formatting rules, see https://plot.ly/r/reference/#scatter-hovertemplate.

```
set.seed(10)
plot_ly(x = rnorm(100, 0, 1e5)) %>%
  add_histogram(
    histnorm = "density",
    hovertemplate = "The height between <br> (%{x}) <br> is %{y:.1e}
    <extra>That's small!</extra>"
  )
```

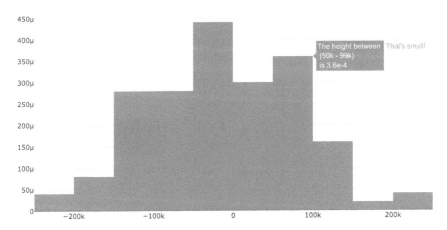

FIGURE 25.5: Using the `hovertemplate` attribute to reference computed variables and their display format inside a custom string.

If you need really specific control over the tooltip, you might consider hiding the tooltip altogether (using `hoverinfo='none'`) and defining your own tooltip. Defining your own tooltip, however, will require knowledge of HTML and JavaScript; see Figure 21.3 for an example of how to display an image on hover instead of a tooltip.

25.2 `ggplotly()` **tooltips**

Similar to how you can use the `text` attribute to supply a custom string in `plot_ly()` (see Section 25.1), you can supply a `text` aesthetic to your **ggplot2** graph, as shown in 25.6:

```
p <- ggplot(mtcars, aes(wt, mpg, text = row.names(mtcars))) +
  geom_point()
ggplotly(p)
```

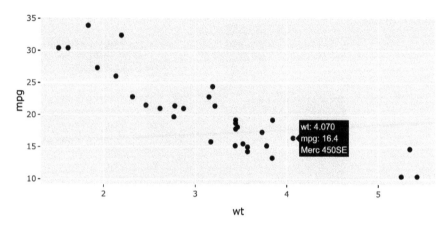

FIGURE 25.6: Using the `text` **ggplot2** aesthetic to supply custom tooltip text to a scatterplot.

By default, `ggplotly()` will display all relevant aesthetic mappings (or computed values), but you can restrict what aesthetics are used to populate the tooltip, as shown in Figure 25.7:

```
ggplotly(p, tooltip = "text")
```

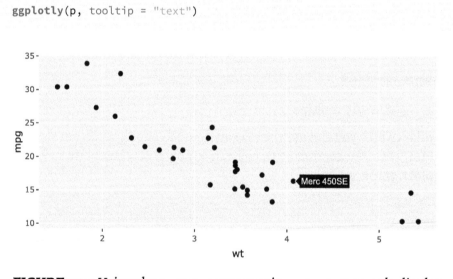

FIGURE 25.7: Using the `tooltip` argument in `ggplotly()` to only display the `text` aesthetic.

When constructing the text to display, ggplotly() runs format() on the computed values. Since some parameters of the format() function can be controlled through global options(), you can use these options() to control the displayed text. This includes the digits option for controlling the number of significant digits used for numerical values as well as scipen for setting a penalty for deciding whether scientific or fixed notation is used for displaying. Figure 25.8 shows how you can temporarily set these options (i.e., avoid altering of your global environment) using the **withr** package (Hester et al., 2018).

```
library(withr)
p <- ggplot(faithfuld, aes(waiting, eruptions)) +
    geom_raster(aes(fill = density))
subplot(
  with_options(list(digits = 1), ggplotly(p)),
  with_options(list(digits = 6, scipen = 20), ggplotly(p))
)
```

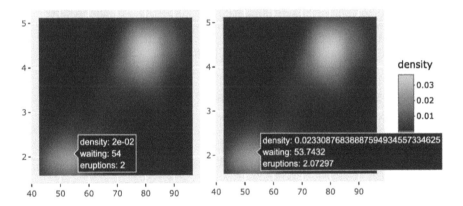

FIGURE 25.8: Leveraging global R options for controlling the displayed values in a ggplotly() tooltip.

These global options are nice for specifying significant/scientific notation, but what about more sophisticated formatting? Sometimes a clever use of the text aesthetic provides a sufficient workaround. Specifically, as Figure 25.9 shows, if one wanted to control a displayed aesthetic value (e.g., y), one could generate a custom string from that variable and supply it to text, then essentially replace text for y in the tooltip:

```
library(scales)
p <- ggplot(txhousing, aes(date, median)) +
  geom_line(aes(
    group = city,
    text = paste("median:", number_si(median))
  ))
ggplotly(p, tooltip = c("text", "x", "city"))
```

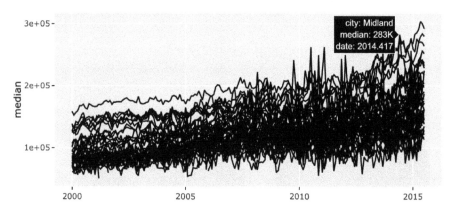

FIGURE 25.9: Using the text aesthetic to replace an auto-generated aesthetic (y).

The approach depicted in Figure 25.9 works for computed values that pertain to raw data values, but what about sophisticated formatting of a summary statistics generated by **ggplot2**? In this case, you'll have to use the return value of ggplotly() which, remember, is a **plotly** object that conforms to the plotly.js spec. That means you can identify trace attribute(s) that contain relevant info (**note:** the plotly_json() function is incredibly for helping to find that information), then use that info to populate a text attribute. Figure 25.10 applies this technique to customize the text that appears when hovering over a geom_smooth() line.

```
# Add a smooth to the previous figure and convert to plotly
w <- ggplotly(p + geom_smooth(se = FALSE))

# This plotly object has two traces: one for
# the raw time series and one for the smooth.
# Try using `plotly_json(w)` to confirm the 2nd
# trace is the smooth line.
length(w$x$data)

# use the `y` attribute of the smooth line
# to generate a custom string (to appear in tooltip)
text_y <- number_si(
  w$x$data[[2]]$y,
  prefix = "Typical median house price: $"
)

# suppress the tooltip on the raw time series
# and supply custom text to the smooth line
w %>%
  style(hoverinfo = "skip", traces = 1) %>%
  style(text = text_y, traces = 2)
```

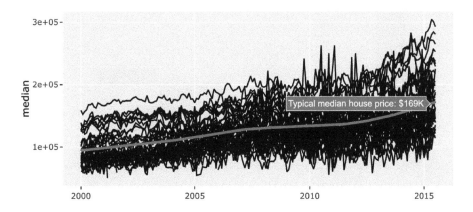

FIGURE 25.10: Using the return value of `ggplotly()` to populate a custom `text` attribute.

25.3 Styling

There is currently one main attribute for controlling the style of a tooltip: `hoverlabel`. With this attribute you can currently set the background color (`bgcolor`), border color (`bordercolor`), and font family/size/color. Figure 25.11 demonstrates how to use it with `plot_ly()` (basically any chart type you use should support it):

```
font <- list(
  family = "Roboto Condensed",
  size = 15,
  color = "white"
)
label <- list(
  bgcolor = "#232F34",
  bordercolor = "transparent",
  font = font
)
plot_ly(x = iris$Petal.Length, hoverlabel = label)
```

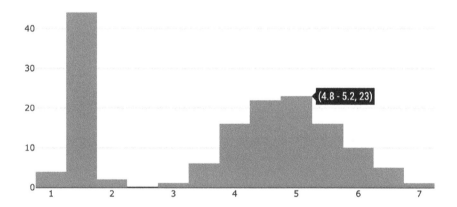

FIGURE 25.11: Using the `hoverlabel` attribute to customize the color and font of the tooltip.

On the other hand, when using ggplotly(), you have to modify the hoverlabel attribute via style() as shown in Figure 25.12

```
qplot(x = Petal.Length, data = iris) %>%
  ggplotly() %>%
  style(hoverlabel = label, marker.color = "#232F34") %>%
  layout(font = font)
```

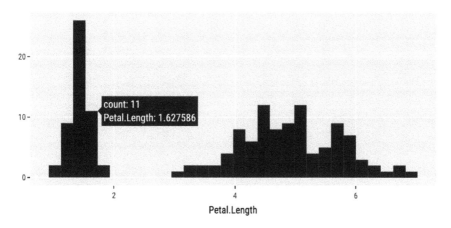

FIGURE 25.12: Using the hoverlabel attribute with ggplotly().

As shown in Sections 25.1 and 25.2, the approach to customize the actual text of a tooltip is slightly different depending on whether you're using ggplotly() or plot_ly(), but styling the appearance of the tooltip is more or less the same in either approach.

26

Control the modebar

By default, the modebar appears in the top right-hand side of a **plotly** graph on mouse hover, and can lead to poor user-experience on small displays. Fortunately, the modebar can be completely customized via the `config()` function. The `config()` function can be helpful for a lot of things: language support (Chapter 30), enabling mathjax (Chapter 31), suppressing tip notifications (e.g., `showTips`), when to scroll on zoom, etc. However, this topic is all about options related to the modebar. To see a complete list of `config()` options, as well as their description, see the config section of the plotly.js `schema()`.

26.1 Remove the entire modebar

The `displayModeBar` option makes it quick and easy to remove the entire modebar.

```
plot_ly() %>%
  config(displayModeBar = FALSE)
```

26.2 Remove the plotly logo

The `displaylogo` option makes it quick and easy to remove the entire modebar.

```
plot_ly() %>%
  config(displaylogo = FALSE)
```

FIGURE 26.1: Removing the plotly logo from the modebar.

26.3   Remove modebar buttons by name

Any modebar buttons can be removed by name via `modeBarButtonsToRemove`. Figure 26.2 demonstrates removal of the 2D zoom in and out button. The full list of modebar buttons included by default can be found at https://github.com/plotly/plotly.js/blob/master/src/components/modebar/buttons.js

```
plot_ly() %>%
  config(modeBarButtonsToRemove = c("zoomIn2d", "zoomOut2d"))
```

FIGURE 26.2: Removing the 'zoomIn2d' and 'zoomOut2d' modebar buttons by name.

26.4   Add custom modebar buttons

It is possible to supply your own modebar button icon that triggers a custom JavaScript function when clicked. You must provide a `name` for the icon and either a SVG path (with just the d attribute) or a full SVG element (to svg). Nowadays, there are a number of free websites that allow you to search icons and download their corresponding SVG information. When supplying path, as in Figure 26.3, you can also define an SVG transform to help size and position the icon. To define a JavaScript function to call upon clicking the icon, you can provide a string to htmlwidgets::JS(). The interactive version of Figure 26.3 adds on-graph text every time the octocat icon is clicked. To learn more about how to leverage JavaScript from R, see Chapter 18.

```r
data(octocat_svg_path, package = "plotlyBook")

octocat <- list(
  name = "octocat",
  icon = list(
    path = octocat_svg_path,
    transform = 'matrix(1 0 0 1 -2 -2) scale(0.7)'
  ),
  click = htmlwidgets::JS(
    "function(gd) {
        var txt = {x: [1], y: [1], text: 'Octocat!', mode: 'text'};
        Plotly.addTraces(gd, txt);
    }"
  )
)

plot_ly() %>%
  config(modeBarButtonsToAdd = list(octocat))
```

FIGURE 26.3: Supplying a custom modebar button with custom behavior.

Note that you can also use modeBarButtons to completely specify which buttons to include in the modebar. As shown in Figure 26.4, with this option, you can supply existing button names and/or your own custom buttons:

```
plot_ly() %>%
  config(modeBarButtons = list(list("zoomIn2d"), list(octocat)))
```

FIGURE 26.4: Specifying the full list of modebar buttons.

26.5 Control image downloads

By default, the toImage modebar button downloads a PNG file using the current size and state of the graph. With toImageButtonOptions, one can specify different sizes and filetypes, which is particularly useful for obtaining a static PDF/WebP/JPG/etc. image of the plot *after* components have been directly manipulated, as leveraged in Figure 12.1. Here's a basic example of configuring the 'toImage' button to download an SVG file that's 200 x 100 pixels:

```
plot_ly() %>%
  config(
    toImageButtonOptions = list(
      format = "svg",
      width = 200,
      height = 100
```

```
  )
)
```

After downloading the SVG file, you can convert it to PDF using the `rsvg_pdf()` function from the **rsvg** package (Ooms, 2018).

27

Working with colors

The JavaScript library underlying **plotly** (plotly.js) has its own support for specifying colors, which is different from how R specifies colors. It currently supports:

- hex (e.g., `"#FF0000"`)
- rgb (e.g., `"rgb(255, 0, 0)"`)
- rgba (e.g., `"rgba(255, 0, 0, 1)"`)
- hsl (e.g., `'hsl(0, 100%, 50%)'`)
- hsv (e.g., `'hsv(0, 100%, 50%)'`)
- Named CSS3 colors `http://www.w3.org/TR/css3-color/#svg-color`

If you use `plot_ly()` and directly specify a plotly.js color attribute (e.g., `marker.color`), you can use any of these formats. Figure 27.1 uses the hsl format:

```
plot_ly(
  x = iris$Petal.Length,
  marker = list(color = "hsl(0, 100%, 50%)")
)
```

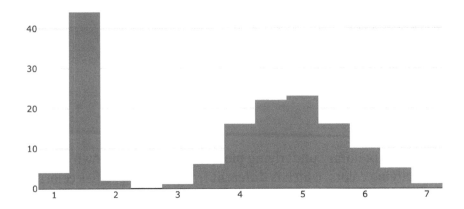

FIGURE 27.1: Specifying a color in plotly.js's supported format.

If you're doing something specific to R, like using ggplotly() and/or the top-level color/colors/stroke/strokes argument in plot_ly(), you'll need to be careful about specifying colors in a way that R and **plotly** can understand. For example, at least currently, you can't specify an hsl string in this way:

```
plot_ly(x = 1, y = 1, color = I("hsl(0, 100%, 50%)"))
```

```
#> Error in grDevices::col2rgb(x, alpha = TRUE):
#>     invalid color name 'hsl(0, 100%, 50%)'
```

Just like in **ggplot2**, you'll have to specify a color in one of the following ways:

- A hexadecimal string of the form "#rrggbb" or "#rrggbbaa".
- Named colors (e.g., "blue"). All supported names are listed in colors().
- An NA for transparent.

This doesn't imply that you can't work in other colorspaces though (e.g., rgb, rgba, hsl, or hsl). The **colorspace** package provides a nice way to create colors in any of these colorspaces and provides a hex() function that you can use to convert any color to a hexadecimal format (Ihaka et al., 2019).

```
library(colorspace)
red <- hex(HLS(0, 0.5, 1))
plot_ly(x = iris$Petal.Length, color = I(red))
```

If you'd like to see more examples of specifying colors, see Chapter 3.

28

Working with symbols and glyphs

The JavaScript library underlying **plotly** (plotly.js) has its own special support for specifying marker symbols by name. As Figure 28.1 shows, there are currently many acceptable marker.symbol values, and all the acceptable values can be accessed through plotly.js's schema().

```r
vals <- schema(F)$traces$scatter$attributes$marker$symbol$values
vals <- grep("-", vals, value = T)
plot_ly() %>%
  add_markers(
    x = rep(1:12, each = 11, length.out = length(vals)),
    y = rep(1:11, times = 12, length.out = length(vals)),
    text = vals,
    hoverinfo = "text",
    marker = list(
      symbol = vals,
      size = 30,
      line = list(
        color = "black",
        width = 2
      )
    )
  )
```

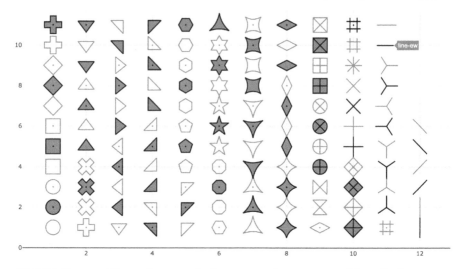

FIGURE 28.1: All marker symbols currently supported by **plotly**.

In addition to these marker symbols, you can also use add_text() to encode data with on-graph text. Moreover, the add_text() function (i.e., a scatter trace with mode="markers") enjoys a lot of the same properties as add_markers() (i.e., a scatter trace with mode="text"). As Figure 28.2 shows, similar to how we can supply typographical glyphs and/or unicode in a custom tooltip, you can supply a character vector of similar content to add_text() (i.e., a scatter trace with mode='text') which renders on-graph text. Furthermore, when using text to render on-graph text, one can leverage the hovertext attribute to display some different text on hover.

```
plot_ly() %>%
  add_text(
    x = rep(2, 2),
    y = 1:2,
    size = I(15),
    text = c(
      "Glyphs: , ♩♫♪ ♫♬",
      "Unicode: \U00AE \U00B6 \U00BF"
    ),
    hovertext = c(
```

```
        "glyphs",
        "unicode"
      ),
      textposition = "left center",
      hoverinfo = "text"
    )
```

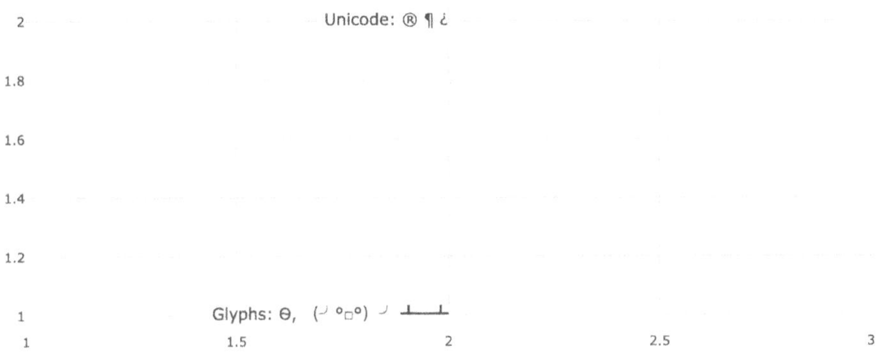

FIGURE 28.2: Using `add_text()` to render on-graph text with typographical glyphs and/or unicode.

Having the ability to encode data with unicode means that we have a virtually endless number of ways to encode data in symbols/glyphs. Just for fun, Figure 28.3 demonstrates how you could plot all the activity emojis using the **emo** package and display the name of the emoji on hover (Wickham et al., 2018).

```
library(emo)
set.seed(100)
jis %>%
  filter(group == "Activities") %>%
  plot_ly(x = runif(nrow(.)), y = runif(nrow(.))) %>%
  add_text(
```

```
    text = ~emoji,
    hovertext = ~name,
    hoverinfo = "text",
    size = I(20)
)
```

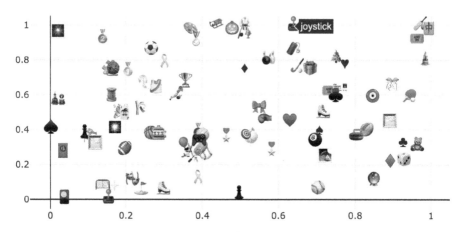

FIGURE 28.3: Using `add_text()` to plot all the activity emojis and leveraging `hovertext` to place the emoji names in the tooltip text.

29

Embedding images

There are a few ways to embed images in a **plotly** graph. Perhaps the easiest is to point the image source to a hyperlink containing a bitmap file (e.g., PNG, JPG, etc.).

```
plot_ly() %>%
  layout(
    images = list(
      source = "https://www.rstudio.com/wp-content/uploads/2018/10/
      RStudio-Logo-Flat.png",
      x = 0, y = 1,
      sizex = 0.2, sizey = 0.1,
      xref = "paper", yref = "paper",
      xanchor = "left", yanchor = "bottom"
    ),
    margin = list(t = 50)
  )
```

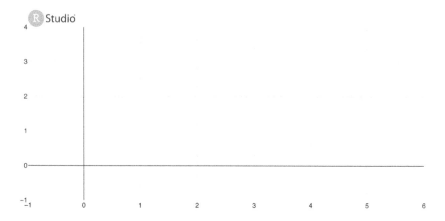

FIGURE 29.1: Embedding an image using a hyperlink. This example uses paper coordinates (i.e., normalized 0-1 scale relative to the graph area) to place the image just above the graph.

The approach in Figure 29.1 has a downside though: if that hyperlink breaks, then so does your plot. It'd be better to download the file to your machine and use the dataURI() function from the **base64enc** package (or similar) to embed the image as a data URI as done in Figure 29.2 (Urbanek, 2015):

```
plot_ly() %>%
  layout(
    images = list(
      source = base64enc::dataURI(file = "images/rstudio.png"),
      x = 0, y = 1,
      sizex = 0.2, sizey = 0.1,
      xref = "paper", yref = "paper",
      xanchor = "left", yanchor = "bottom"
    ),
    margin = list(t = 50)
  )
```

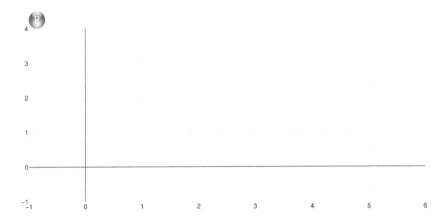

FIGURE 29.2: Embedding an image using a data URI. This approach ensures your image stays embedded in the plot.

Another approach is to convert a raster object into a data URI, which the `raster2uri()` function in **plotly** is designed to do. R actually ships with native support for raster objects and many image processing R packages either build on this data structure or provide a utility to convert to a raster object (perhaps via `as.raster()`). As done in Figure 29.3, the `readPNG()` function from the **png** package reads image data in an R array, which can be converted to a raster object (Urbanek, 2013).

```
pen <- png::readPNG("images/penguin.png")

plot_ly() %>%
  layout(
    images = list(
      source = raster2uri(as.raster(pen)),
      x = 2, y = 2,
      sizex = 2, sizey = 1,
      xref = "x", yref = "y",
      xanchor = "left", yanchor = "bottom",
      sizing = "stretch"
    )
  )
```

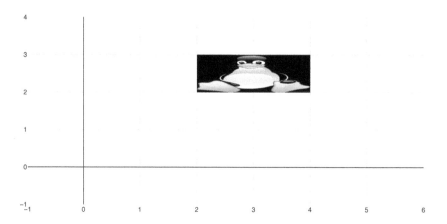

FIGURE 29.3: Reading a PNG image into R with `png::readPNG()`, converting the RBGA array into a raster object, generating a data URI of the raster via `raster2uri()`, then embedding the URI as an image in a **plotly** graph.

Moreover, thanks to the R package **magick**, it's possible to rasterize non-raster file formats (e.g., PDF, SVG, etc.) directly in R, so if you wanted, you could also embed non-bitmap images by using `image_read()` and `image_convert()` to generate a raster object (Ooms, 2019).

30

Language support

The `locale` argument of the `config()` function allows one to render on-graph text using another language. Figure 30.1 shows how setting `locale='ja'` will render text in Japanese.

```r
today <- Sys.Date()
x <- seq.Date(today, today + 360, by = "day")
plot_ly(x = x, y = rnorm(length(x))) %>%
  add_lines() %>%
  config(locale = "ja")
```

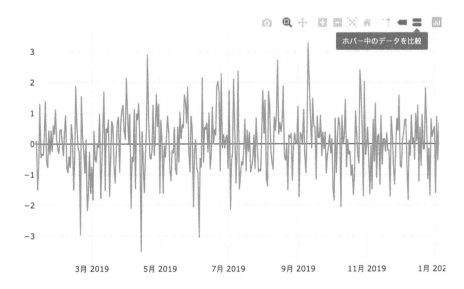

FIGURE 30.1: Using the `locale` argument of the `config()` function to render on-graph text in another language.

Table 30.1 contains a list of all the supported locale codes and the language they correspond to.

TABLE 30.1: Language support in plotly. To use one of these languages, supply the locale code to the `locale` argument in the `config()` function.

Language	Code
Afrikaans	af
Albanian	sq
Amharic	am
Arabic - Algeria	ar
Arabic - Bahrain	ar
Arabic - Egypt	ar
Arabic - Iraq	ar
Arabic - Jordan	ar
Arabic - Kuwait	ar
Arabic - Lebanon	ar
Arabic - Libya	ar
Arabic - Morocco	ar
Arabic - Oman	ar
Arabic - Qatar	ar
Arabic - Saudi Arabia	ar
Arabic - Syria	ar
Arabic - Tunisia	ar
Arabic - United Arab Emirates	ar
Arabic - Yemen	ar
Armenian	hy
Azeri - Cyrillic	az
Azeri - Latin	az
Basque	eu
Bosnian	bs
Bulgarian	bg
Catalan	ca

TABLE 30.1: Language support in plotly. To use one of these languages, supply the locale code to the `locale` argument in the `config()` function. *(continued)*

Language	Code
Croatian	hr
Czech	cs
Danish	da
Dutch - Belgium	nl
Dutch - Netherlands	nl
Estonian	et
FYRO Macedonia	mk
Faroese	fo
Farsi - Persian	fa
Finnish	fi
French - Belgium	fr
French - Cameroon	fr
French - Canada	fr
French - Congo	fr
French - Cote d'Ivoire	fr
French - France	fr
French - Luxembourg	fr
French - Mali	fr
French - Monaco	fr
French - Morocco	fr
French - Senegal	fr
French - Switzerland	fr
French - West Indies	fr
Galician	gl
Georgian	ka
German - Austria	de
German - Germany	de
German - Liechtenstein	de
German - Luxembourg	de

TABLE 30.1: Language support in plotly. To use one of these languages, supply the locale code to the `locale` argument in the `config()` function. (*continued*)

Language	Code
German - Switzerland	de
Greek	el
Gujarati	gu
Hebrew	he
Hungarian	hu
Icelandic	is
Indonesian	id
Italian - Italy	it
Italian - Switzerland	it
Japanese	ja
Khmer	km
Korean	ko
Latvian	lv
Lithuanian	lt
Malay - Brunei	ms
Malay - Malaysia	ms
Malayalam	ml
Maltese	mt
Polish	pl
Punjabi	pa
Raeto-Romance	rm
Romanian - Moldova	ro
Romanian - Romania	ro
Russian	ru
Russian - Moldova	ru
Serbian - Cyrillic	sr
Serbian - Latin	sr
Slovak	sk
Slovenian	sl

TABLE 30.1: Language support in plotly. To use one of these languages, supply the locale code to the `locale` argument in the `config()` function. (*continued*)

Language	Code
Spanish - Argentina	es
Spanish - Bolivia	es
Spanish - Chile	es
Spanish - Colombia	es
Spanish - Costa Rica	es
Spanish - Dominican Republic	es
Spanish - Ecuador	es
Spanish - El Salvador	es
Spanish - Guatemala	es
Spanish - Honduras	es
Spanish - Mexico	es
Spanish - Nicaragua	es
Spanish - Panama	es
Spanish - Paraguay	es
Spanish - Peru	es
Spanish - Puerto Rico	es
Spanish - Spain (Traditional)	es
Spanish - Uruguay	es
Spanish - Venezuela	es
Swahili	sw
Swedish - Finland	sv
Swedish - Sweden	sv
Tamil	ta
Tatar	tt
Thai	th
Turkish	tr
Ukrainian	uk
Urdu	ur
Vietnamese	vi
Welsh	cy

31

LaTeX rendering

LaTeX rendering via MathJax is possible via the `TeX()` function which flags a character vector as LaTeX. To load MathJaX externally (meaning an internet connection is needed for TeX rendering), set the new `mathjax` argument in `config()` to `"cdn"`. Figure 31.1 demonstrates how to render LaTeX in the plot and axis titles.

```r
library(plotly)
data(co2, package = "datasets")

plot_ly() %>%
  add_lines(x = zoo::index(co2), y = co2) %>%
  layout(
    title = TeX("CO_2 \\text{measured in } \\frac{parts}{million}"),
    xaxis = list(title = "Time"),
    yaxis = list(
      title = TeX("\\text{Atmospheric concentration of CO}_2")
    )
  ) %>%
  config(mathjax = "cdn")
```

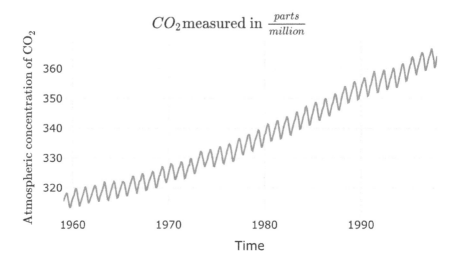

FIGURE 31.1: Rendering LaTeX in the plot and axis titles.

Figure 31.2 demonstrates how to render LaTeX with on-graph text. There are two ways to draw on-graph text: `add_text()` which is a scatter trace with a mode of text and `add_annotations()` which is part of the graph's layout. The main difference is that `add_text()` is able to display tooltips and `add_annotations()` is able to display arrows.

```
plotly_empty(showlegend = FALSE, hoverinfo = "x+y") %>%
  add_annotations(
    x = 1, y = 2,
    showarrow = FALSE,
    text = TeX(
      "\\text{The sample mean:} \\sum_{i=1}^n x_i \\text{ where}"
    )
  ) %>%
  add_text(
    x = 1, y = 1, size = I(100),
    text = TeX("x_i \\sim N(\\mu, \\sigma)")
  ) %>%
  add_annotations(x = 1, y = 0, text = TeX("E[x_i]")) %>%
  add_text(
```

```
  x = 1, y = 0, text = TeX("\\mu"), textposition = "bottom"
) %>%
config(mathjax = "cdn")
```

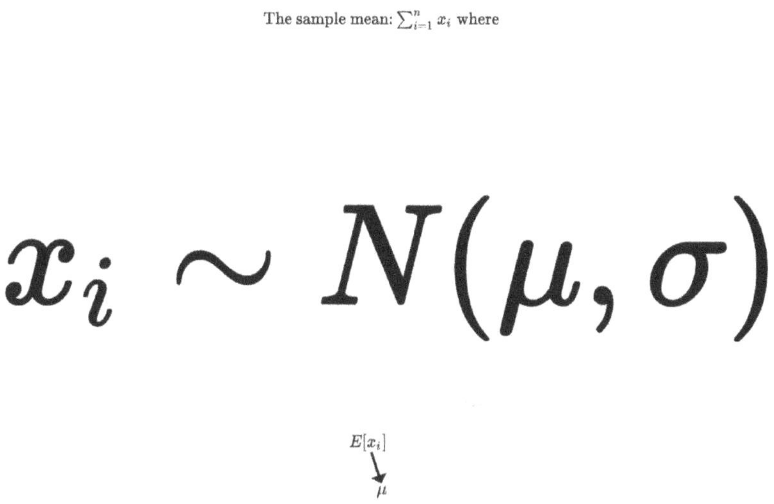

The sample mean: $\sum_{i=1}^{n} x_i$ where

$$x_i \sim N(\mu, \sigma)$$

$E[x_i]$

μ

FIGURE 31.2: Rendering LaTeX using `add_text()` and `add_annotations()`.

To use a local version of MathJax (so that your graphs will render without an internet connection), you need to inform **plotly** where it's located. If you don't already have MathJax locally, I recommend downloading the official MathJax git repo. Here's how to do that using terminal commands:

```
$ git clone https://github.com/mathjax/MathJax.git
$ cd MathJax
```

Now set the `PLOTLY_MATHJAX_PATH` environment variable so that **plotly** knows where that MathJax folder lives. I recommend setting this variable in you `.Rprofile` so you don't have to reset it every time you restart R:

```
$ export PLOTLY_MATHJAX_PATH=`pwd`
```

```
$ echo "Sys.setenv('PLOTLY_MATHJAX_PATH' = '$PLOTLY_MATHJAX_PATH')" >>
+    ~/.Rprofile
```

Finally, once `PLOTLY_MATHJAX_PATH` is set, specify `mathjax="local"` in `config()`:

```
config(last_plot(), mathjax = "local")
```

___

31.1   MathJax caveats

1. MathJax rendering in tooltips currently isn't supported[1].

2. At least currently, plotly.js requires SVG-based rendering which doesn't play nicely with HTML-based rendering (e.g., **rmarkdown** documents and **shiny** apps) . If you need both the SVG and HTML rendering, consider `<iframe>`-ing your plotly graph(s) into the larger document (see here[2] for an example).

3. Due to the size and nature of MathJax, using `htmlwidget::saveWidget()` with `selfcontained = TRUE` won't work. At least for now, when you need to save a plotly graph (p) with local MathJax, do `htmlwidget::saveWidget(p, selfcontained = FALSE)`

___

[1]https://github.com/plotly/plotly.js/issues/559
[2]https://github.com/ropensci/plotly/blob/master/inst/examples/rmd/MathJax/index.Rmd

32

The data-plot-pipeline

As Section 2.1 first introduced, we can express multi-layer **plotly** graphs as a sequence (or, more specifically, a directed acyclic graph) of **dplyr** data manipulations and mappings to visuals. For example, to create Figure 32.1, we could group txhousing by city to ensure the first layer of add_lines() draws a different line for each city, then filter() down to Houston so that the second call to add_lines() draws only Houston.

```
allCities <- txhousing %>%
  group_by(city) %>%
  plot_ly(x = ~date, y = ~median) %>%
  add_lines(alpha = 0.2, name = "Texan Cities", hoverinfo = "none")

allCities %>%
  filter(city == "Houston") %>%
  add_lines(name = "Houston")
```

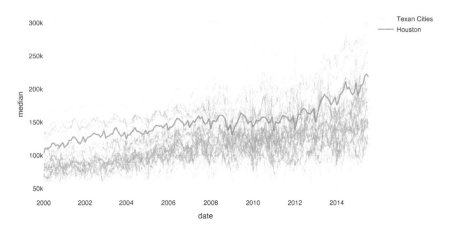

FIGURE 32.1: Monthly median house prices over time for 46 Texan cities (in blue). Houston is highlighted in orange.

Sometimes, the directed acyclic graph property of a **magrittr** pipeline can be too restrictive for certain types of plots. In this example, after filtering the data down to Houston, there is no way to recover the original data inside the pipeline. The add_fun() function helps to workaround this restriction[1]; it works by applying a function to the plotly object, but does not affect the data associated with the plotly object. This effectively provides a way to isolate data transformations within the pipeline[2]. Figure 32.2 uses this idea to highlight both Houston and San Antonio.

```
allCities %>%
  add_fun(function(plot) {
    plot %>% filter(city == "Houston") %>%
      add_lines(name = "Houston")
  }) %>%
  add_fun(function(plot) {
    plot %>% filter(city == "San Antonio") %>%
```

---

[1]Credit to Winston Chang and Hadley Wickham for this idea. The add_fun() is very much like layer_f() function in **ggvis**.

[2]Effectively putting a pipeline inside a pipeline

```
    add_lines(name = "San Antonio")
})
```

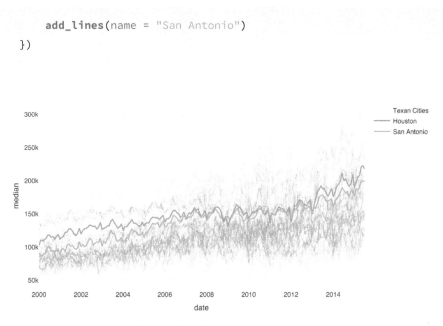

FIGURE 32.2: Monthly median house price in Houston and San Antonio in comparison to other Texan cities.

It is useful to think of the function supplied to add_fun() as a "layer" function; a function that accepts a plot object as input, possibly applies a transformation to the data, and maps that data to visual objects. To make layering functions more modular, flexible, and expressive, the add_fun() allows you to pass additional arguments to a layer function. Figure 32.3 makes use of this pattern, by creating a reusable function for layering both a particular city as well as the first, second, and third quartile of median monthly house sales (by city).

```
# reusable function for highlighting a particular city
layer_city <- function(plot, name) {
  plot %>% filter(city == name) %>% add_lines(name = name)
}

# reusable function for plotting overall median & IQR
layer_iqr <- function(plot) {
  plot %>%
```

```
    group_by(date) %>%
    summarise(
      q1 = quantile(median, 0.25, na.rm = TRUE),
      m = median(median, na.rm = TRUE),
      q3 = quantile(median, 0.75, na.rm = TRUE)
    ) %>%
    add_lines(y = ~m, name = "median", color = I("black")) %>%
    add_ribbons(
      ymin = ~q1, ymax = ~q3,
      name = "IQR", color = I("black")
    )
}


allCities %>%
  add_fun(layer_iqr) %>%
  add_fun(layer_city, "Houston") %>%
  add_fun(layer_city, "San Antonio")
```

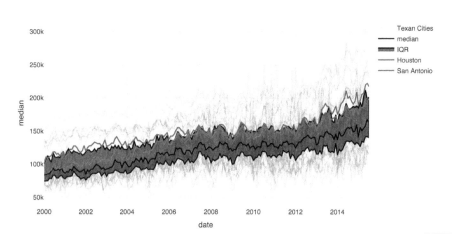

FIGURE 32.3: First, second, and third quartile of median monthly house price in Texas.

A layering function does not have to be a data-plot-pipeline itself. Its only requirement on a layering function is that the first argument is a plot object and it returns a plot object. This provides an opportunity to say, fit a model to the plot data, extract the model components you desire, and map those components to visuals. Furthermore, since **plotly**'s add_*() functions don't require a data.frame, you can supply those components directly to attributes (as long as they are well defined), as done in Figure 32.4 via the **forecast** package (Hyndman, 2018).

```
library(forecast)
layer_forecast <- function(plot) {
  d <- plotly_data(plot)
  series <- with(d,
    ts(median, frequency = 12, start = c(2000, 1), end = c(2015, 7))
  )
  fore <- forecast(ets(series), h = 48, level = c(80, 95))
  plot %>%
    add_ribbons(x = time(fore$mean), ymin = fore$lower[, 2],
                ymax = fore$upper[, 2], color = I("gray95"),
                name = "95% confidence", inherit = FALSE) %>%
    add_ribbons(x = time(fore$mean), ymin = fore$lower[, 1],
                ymax = fore$upper[, 1], color = I("gray80"),
                name = "80% confidence", inherit = FALSE) %>%
    add_lines(x = time(fore$mean), y = fore$mean, color = I("blue"),
              name = "prediction")
}

txhousing %>%
  group_by(city) %>%
  plot_ly(x = ~date, y = ~median) %>%
  add_lines(alpha = 0.2, name = "Texan Cities", hoverinfo="none") %>%
  add_fun(layer_iqr) %>%
  add_fun(layer_forecast)
```

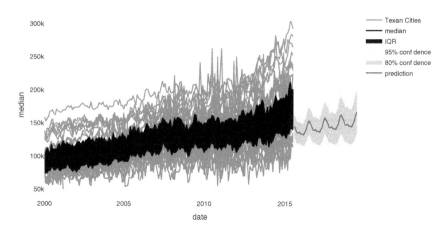

FIGURE 32.4: Layering on a 4-year forecast from an exponential smoothing state space model.

In summary, the "data-plot-pipeline" is desirable for a number of reasons: (1) makes your code easier to read and understand, (2) encourages you to think of both your data and plots using a single, uniform data structure, which (3) makes it easy to combine and reuse transformations.

33

Improving `ggplotly()`

Since the `ggplotly()` function returns a **plotly** object, we can use that object in the same way you can use any other **plotly** object. Modifying this object is *always* going to be useful when you want more control over certain (interactive) behavior that **ggplot2** doesn't provide an API to describe[1], for example:

- `layout()` for modifying aspects of the layout, which can be used to do many things, for example:
    - Change default `hovermode`[2] behavior (See Figure 33.1).
    - Stylizing hover labels (`hoverlabel`[3]).
    - Changing click+drag mode (`dragmode`[4]) and/or constraining rectangular selections (`dragmode='select'`) vertically or horizontally (`selectdirection`[5]).
    - Add dropdowns https://plot.ly/r/dropdowns/, sliders https://plot.ly/r/sliders/, and rangesliders (see Figure 33.1).
- `style()` for modifying data-level attributes, which can be used to:
    - Control the tooltip content and styling (see Section 25.2).
    - Turn hovering on/off (see Figure 33.4).
    - Add marker points to lines (e.g., `style(p, mode = "markers+lines")`).
- `config()` for modifying the plot configuration, which can be used to:
    - Control the modebar (see Chapter 26).
    - Change the default language (see Chapter 30).
    - Enable LaTeX rendering (see Chapter 31).
    - Enable editable shapes (see Chapter 12).

---

[1]It can also be helpful for correcting translations that `ggplotly()` doesn't get quite right.
[2]https://plot.ly/r/reference/#layout-hovermode
[3]https://plot.ly/r/reference/#layout-hoverlabel
[4]https://plot.ly/r/reference/#layout-dragmode
[5]https://plot.ly/r/reference/#layout-selectdirection

In addition to using the functions above to modify ggplotly()'s return value, the ggplotly() function itself provides some arguments for controlling that return value. In this chapter, we'll see a couple of them:

- dynamicTicks: should plotly.js dynamically generate axis tick labels? Dynamic ticks are useful for updating ticks in response to zoom/pan interactions; however, they cannot always reproduce labels as they would appear in the static ggplot2 image (see Figure 33.1).
- layerData: which **ggplot2** layer's data should be returned (see Figure 33.5)?

33.1   Modifying layout

Any aspect of a **plotly** object's layout can be modified[6] via the layout() function. By default, since it doesn't always make sense to compare values, ggplotly() will usually set layout.hovermode='closest'. As shown in Figure 33.1, when we have multiple y-values of interest at a specific x-value, it can be helpful to set layout.hovermode='x'. Moreover, for a long time series graph like this, zooming in on the x-axis can be useful; dynamicTicks allows plotly.js to handle the generation of axis ticks and the rangeslider() allows us to zoom on the x-axis without losing the global context.

```
library(babynames)
nms <- filter(babynames, name %in% c("Sam", "Alex"))
p <- ggplot(nms) +
  geom_line(aes(year, prop, color = sex, linetype = name))

ggplotly(p, dynamicTicks = TRUE) %>%
  rangeslider() %>%
  layout(hovermode = "x")
```

---

[6]Or, in the case of cumulative attributes, like shapes, images, annotations, etc, these items will be added to the existing items.

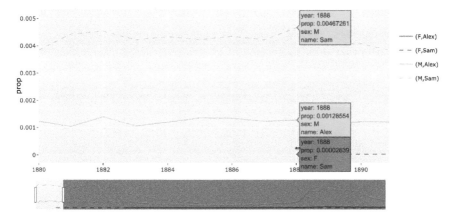

FIGURE 33.1: Adding `dynamicTicks`, a `rangeslider()`, and a comparison hovermode to improve the interactive experience of a `ggplotly()` graph. For a video demonstration of the interactive, see `https://bit.ly/ggplotly-rangeslider`. For the interactive, see `https://plotly-r.com/interactives/ggplotly-rangeslider.html`

Since a single **plotly** object can only have one layout, modifying the layout of `ggplotly()` is fairly easy, but it's trickier to modify the data underlying the graph.

33.2 Modifying data

As mentioned previously, `ggplotly()` translates each ggplot2 layer into one or more plotly.js traces. In this translation, it is forced to make a number of assumptions about trace attribute values that may or may not be appropriate for the use case. To demonstrate, consider Figure 33.2, which shows hover information for the points, the fitted line, and the confidence band. How could we make it so that hover information is only displayed for the points and not for the fitted line and confidence band?

```
p <- ggplot(mtcars, aes(x = wt, y = mpg)) +
    geom_point() + geom_smooth()
ggplotly(p)
```

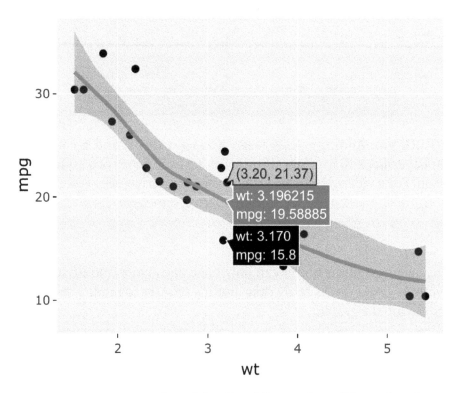

FIGURE 33.2: A scatterplot with a fitted line and confidence band.

The **ggplot2** package doesn't provide an API for interactive features, but by changing the hoverinfo[7] attribute to "none", we can turn off hover for the relevant traces. This sort of task (i.e., modifying trace attribute values) is best achieved through the `style()` function. Before using it, you may want to study the underlying traces with `plotly_json()` which uses the **listviewer** package to display a convenient interactive view of the JSON object sent to plotly.js (de Jong and Russell, 2016). By clicking on the arrow next to the data element, you can see the traces (data)

---

[7]https://plot.ly/r/reference/#scatter-hoverinfo

behind the plot. As shown in Figure 33.3, we have three traces: one for the `geom_point()` layer and two for the `geom_smooth()` layer.

```
plotly_json(p)
```

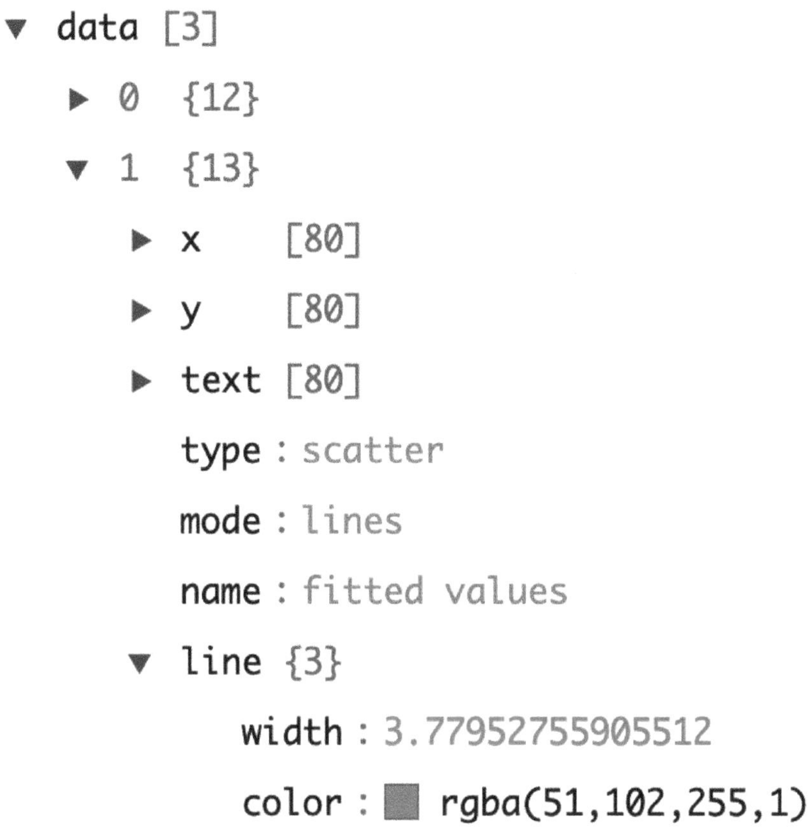

▼ **data** [3]

   ▶ 0 {12}

   ▼ 1 {13}

      ▶ x   [80]

      ▶ y   [80]

      ▶ text [80]

      type : scatter

      mode : lines

      name : fitted values

      ▼ line {3}

         width : 3.77952755905512

         color : ▮ rgba(51,102,255,1)

FIGURE 33.3: Using listviewer to inspect the JSON representation of a plotly object.

This output indicates that the fitted line and confidence band are implemented in the 2nd and 3rd trace of the plotly object, so to turn off the hover of those traces:

```
style(p, hoverinfo = "none", traces = 2:3)
```

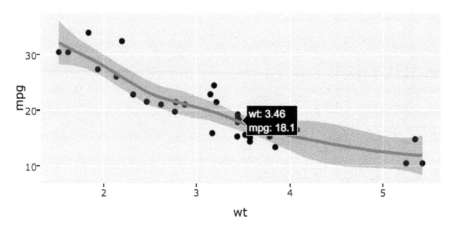

FIGURE 33.4: Using the `style()` function to modify hoverinfo attribute values of a plotly object created via `ggplotly()` (by default, `ggplotly()` displays hoverinfo for all traces). In this case, the hoverinfo for a fitted line and error bounds are hidden.

33.3   Leveraging statistical output

Since `ggplotly()` returns a plotly object, and **plotly** objects can have data attached to them, it attaches data from **ggplot2** layer(s) (either before or after summary statistics have been applied). Furthermore, since each ggplot layer owns a data frame, it is useful to have some way to specify the particular layer of data of interest, which is done via the `layerData` argument in `ggplotly()`. Also, when a particular layer applies a summary statistic (e.g., `geom_bin()`), or applies a statistical model (e.g., `geom_smooth()`) to the data, it might be useful to access the output of that transformation, which is the point of the `originalData` argument in `ggplotly()`.

```
p <- ggplot(mtcars, aes(x = wt, y = mpg)) +
  geom_point() + geom_smooth()
p %>%
  ggplotly(layerData = 2, originalData = FALSE) %>%
  plotly_data()
#> # A tibble: 80 x 14
#>        x     y   ymin  ymax    se flipped_aes PANEL group
#>    <dbl> <dbl> <dbl> <dbl> <dbl> <lgl>        <fct> <int>
#> 1   1.51  32.1  28.1  36.0  1.92 FALSE            1    -1
#> 2   1.56  31.7  28.2  35.2  1.72 FALSE            1    -1
#> 3   1.61  31.3  28.1  34.5  1.54 FALSE            1    -1
#> 4   1.66  30.9  28.0  33.7  1.39 FALSE            1    -1
#> 5   1.71  30.5  27.9  33.0  1.26 FALSE            1    -1
#> 6   1.76  30.0  27.7  32.4  1.16 FALSE            1    -1
#> # ... with 74 more rows, and 6 more variables:
#> #   colour <chr>, fill <chr>, size <dbl>,
#> #   linetype <dbl>, weight <dbl>, alpha <dbl>
```

The data shown above is the data ggplot2 uses to actually draw the fitted values (as a line) and standard error bounds (as a ribbon). Figure 33.5 leverages this data to add additional information about the model fit; in particular, it adds vertical lines and annotations at the x-values that are associated with the highest and lowest amount uncertainty in the fitted values. Producing a plot like this with **ggplot2** would be impossible using geom_smooth() alone.[8] Providing a simple visual clue like this can help combat visual misperceptions of uncertainty bands due to the sine illusion (VanderPlas and Hofmann, 2015).

```
p %>%
  ggplotly(layerData = 2, originalData = FALSE) %>%
  add_fun(function(p) {
    p %>% slice(which.max(se)) %>%
```

---

[8]It could be recreated by fitting the model via loess(), obtaining the fitted values and standard error with predict(), and feeding those results into geom_line()/geom_ribbon()/geom_text()/geom_segment(), but that process is much more onerous.

```
      add_segments(x = ~x, xend = ~x, y = ~ymin, yend = ~ymax) %>%
      add_annotations("Maximum uncertainty", ax = 60)
  }) %>%
  add_fun(function(p) {
    p %>% slice(which.min(se)) %>%
      add_segments(x = ~x, xend = ~x, y = ~ymin, yend = ~ymax) %>%
      add_annotations("Minimum uncertainty")
  })
```

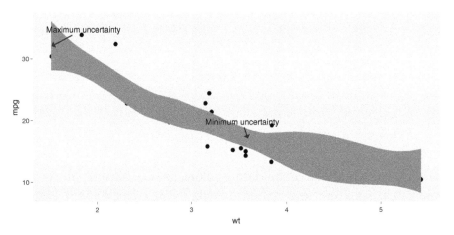

FIGURE 33.5: Leveraging data associated with a `geom_smooth()` layer to display additional information about the model fit.

In addition to leveraging output from `StatSmooth`, it is sometimes useful to leverage output of other statistics, especially for annotation purposes. Figure 33.6 leverages the output of `statBin` to add annotations to a stacked bar chart. Annotation is primarily helpful for displaying the heights of bars in a stacked bar chart, since decoding the heights of bars is a fairly difficult perceptual task (Cleveland and McGill, 1984). As a result, it is much easier to compare bar heights representing the proportion of diamonds with a given clarity across various diamond cuts.

```
p <- ggplot(diamonds, aes(cut, fill = clarity)) +
  geom_bar(position = "fill")

ggplotly(p, originalData = FALSE) %>%
  mutate(ydiff = ymax - ymin) %>%
  add_text(
    x = ~x, y = ~(ymin + ymax) / 2,
    text = ~ifelse(ydiff > 0.02, round(ydiff, 2), ""),
    showlegend = FALSE, hoverinfo = "none",
    color = I("white"), size = I(9)
  )
```

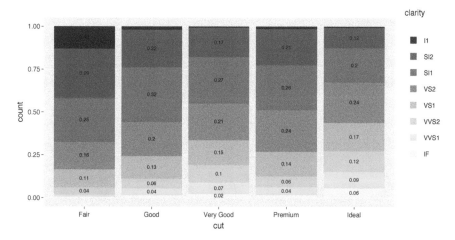

FIGURE 33.6: Leveraging output from `statBin` to add annotations to a stacked bar chart (created via `geom_bar()`) which makes it easier to compare bar heights.

Another useful application is labelling the levels of each piece/polygon output by `statDensity2d` as shown in Figure 33.7. Note that, in this example, the `add_text()` layer takes advantage of `ggplotly()`'s ability to inherit aesthetics from the global mapping. Furthermore, since `originalData` is FALSE, it attaches the "built" aesthetics (i.e., the x/y positions after `statDensity2d` has been applied to the raw data).

```
p <- ggplot(MASS::geyser, aes(x = waiting, y = duration)) +
  geom_density2d()

ggplotly(p, originalData = FALSE) %>%
  group_by(piece) %>%
  slice(which.min(y)) %>%
  add_text(
    text = ~level, size = I(16), color = I("black"), hoverinfo="none"
  )
```

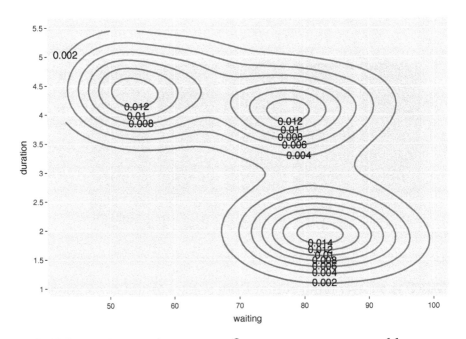

FIGURE 33.7: Leveraging output from `StatDensity2d` to add annotations to contour levels.

34

Translating custom ggplot2 geoms

Version 2.0.0 of **ggplot2** introduced a way for other R packages to implement custom geoms. Some great examples include: **ggrepel**, **ggalt**, **ggraph**, **geomnet**, **ggmosaic** and **ggtern** (Rudis, 2016; Pedersen, 2016; Tyner and Hofmann, 2016; Jeppson et al., 2016; Hamilton, 2016).[1] Although the ggplotly() function translates most of the geoms bundled with the **ggplot2** package, it has no way of knowing about the rendering rules for custom geoms. The **plotly** package does, however, provide 2 generic functions based on the S3 scheme that can be leveraged to inform ggplotly() about these rules (Chambers, 1992).[2] To date, the **ggmosaic** and **ggalt** packages have taken advantage of this infrastructure to provide translations of their custom geoms to plotly.

In **ggplot2**, many geoms are special cases of other geoms. For example, geom_line() is equivalent to geom_path() once the data is sorted by the x variable. For cases like this, when a geom can be reduced to another lower-level (i.e., basic) geom, authors just have to write a method for the to_basic() generic function in **plotly**. In fact, within the package itself, the to_basic() function has a GeomLine method which simply sorts the data by the x variable, then returns it with a class of GeomPath prefixed.

```
getS3method("to_basic", "GeomLine")
#> function (data, prestats_data, layout, params, p, ...)
#> {
#>     data <- data[order(data[["x"]]), ]
```

---

[1]There are many other useful extension packages that are listed on this website – https://www.ggplot2-exts.org

[2]For those new to S3, http://adv-r.had.co.nz/S3.html provides an approachable introduction and overview (Wickham, 2014a).

```
#>       prefix_class(data, "GeomPath")
#> }
#> <bytecode: 0x7f98e9595c70>
#> <environment: namespace:plotly>
```

If you have implemented a custom geom, say GeomCustom, rest assured that the data passed to to_basic() will be of class GeomCustom when ggplotly() is called on a plot with your geom. And assuming GeomCustom may be reduced to another lower-level geom support by plotly, a to_basic.GeomCustom() method that transforms the data into a form suitable for that lower-level geom is sufficient for adding support. Moreover, note that the data passed to to_basic() is essentially the last form of the data *before* the render stage and *after* statistics have been performed. This makes it trivial to add support for geoms like GeomXspline from the **ggalt** package.

```
library(ggalt)
getS3method("to_basic", "GeomXspline")
```

```
function(data, prestats_data, layout, params, p, ...) {
  data <- data[order(data[["x"]]), ]
  prefix_class(data, "GeomPath")
}
```

As shown in Figure 34.1, once the conversion has been provided, users can call ggplotly() on the ggplot object containing the custom geom just like any other ggplot object.

```
# example from `help(geom_xspline)`
set.seed(1492)
dat <- data.frame(
  x = c(1:10, 1:10, 1:10),
  y = c(
    sample(15:30, 10),
    2 * sample(15:30, 10),
    3 * sample(15:30, 10)
```

```
  ),
  group = factor(c(rep(1, 10), rep(2, 10), rep(3, 10)))
)
p <- ggplot(dat, aes(x, y, group = group, color = factor(group))) +
  geom_point(color = "black") +
  geom_smooth(se = FALSE, linetype = "dashed", size = 0.5) +
  geom_xspline(spline_shape = 1, size = 0.5)
ggplotly(p) %>% hide_legend()
```

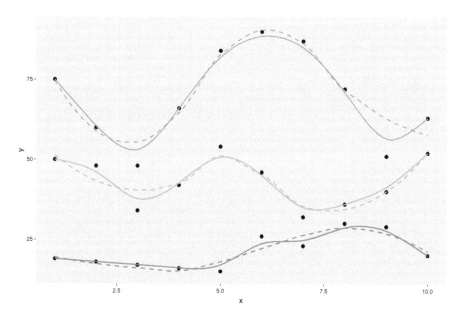

FIGURE 34.1: Converting GeomXspline from the **ggalt** package to plotly.js via ggplotly().

In more complicated cases, where your custom geom cannot be converted to a lower level geom, a custom method for the geom2trace() generic is required (methods(geom2trace) lists all the basic geoms that we natively support). This method should involve a conversion from a data frame to a list-like object conforming to the plotly.js figure reference[3].

---

[3] https://plot.ly/r/reference

Bibliography

Ahlberg, C., Williamson, C., and Shneiderman, B. (1991). Dynamic queries for information exploration: An implementation and evaluation. In *ACM CHI '92 Conference Proceedings*, volume 21, pages 619–626.

Allaire, J. (2016). *flexdashboard: R Markdown Format for Flexible Dashboards*. R package version 0.3.

Asimov, D. (1985). The grand tour: A tool for viewing multidimensional data. *SIAM J. Sci. Stat. Comput.*, 6(1):128–143.

Auguie, B. (2016). *gridExtra: Miscellaneous Functions for Grid Graphics*. R package version 2.2.1.

Bache, S. M. and Wickham, H. (2014). *magrittr: A Forward-Pipe Operator for R*. R package version 1.5.

Becker et al. (1996). The visual design and control of trellis display. *Journal of Computational and Graphical Statistics*, 5(2):123–155.

Berkeley Institute for Data Science (2016). mpl colormaps.

Bostock, M., Ogievetsky, V., and Heer, J. (2011). D3: Data-Driven Documents. *IEEE Trans. Visualization & Comp. Graphics (Proc. InfoVis)*.

Bryan, J. (2015). *gapminder: Data from Gapminder*. R package version 0.2.0.

Buja, A., McDonald, J. A., Michalak, J., and Stuetzle, W. (1991). Interactive data visualization using focusing and linking. *IEEE Proceedings of Visualization*, pages 1–8.

Chamberlain, S. and Teucher, A. (2018). *geojsonio: Convert Data from and to 'GeoJSON' or 'TopoJSON'*. R package version 0.6.0.

Chambers, J. M. (1992). Classes and methods: object-oriented programming in S. In Chambers, J. M. and Hastie, T. J., editors, *Statistical Models in S*. Wadsworth & Brooks/Cole.

Chang, W. (2016). *webshot: Take Screenshots of Web Pages*. R package version 0.3.2.

Chang, W. (2017). Interactive plots (in shiny). http://shiny.rstudio.com/articles/plot-interaction.html.

Chang, W. and Borges Ribeiro, B. (2018). *shinydashboard: Create Dashboards with 'Shiny'*. R package version 0.7.1.

Chang, W. and Luraschi, J. (2018). *profvis: Interactive Visualizations for Profiling R Code*. R package version 0.3.5.

Cheng, J. (2018a). Case study: converting a shiny app to async. https://rstudio.github.io/promises/articles/casestudy.html.

Cheng, J. (2018b). *promises: Abstractions for Promise-Based Asynchronous Programming*. R package version 1.0.1.

Cheng, J. (2018c). Using leaflet with shiny. https://rstudio.github.io/leaflet/shiny.html.

Cleveland, W. S. and Hafen, R. (2014). Divide and recombine: Data science for large complex data. *Statistical Analysis and Data Mining: The ASA Data Science Journal*, 7:425–433.

Cleveland, W. S. and McGill, R. (1984). Graphical Perception: Theory, Experimentation, and Application to the Development of Graphical Methods. *Journal of the American Statistical Association*, 79:531–554.

Cook, D., Buja, A., and Swayne, D. F. (2007). Interactive High-Dimensional Data Visualization. *Journal of Computational and Graphical Statistics*, pages 1–23.

Cook, D., Ebert, A., Hofmann, H., Hyndman, R., Lumley, T., Marwick, B., Sievert, C., Sun, M., Talagala, D., Tierney, N., Tomasetti, N., Wang, E., and Zhou, F. (2017). *eechidna: Exploring Election and Census Highly Informative Data Nationally for Australia*. R package version 1.1.

Cook, D. and Swayne, D. F. (2007). *Interactive and dynamic graphics for data analysis : with R and GGobi.* Use R ! Springer, New York.

Curley, J. (2016). *engsoccerdata: English and European Soccer Results 1871-2016.* R package version 0.1.5.

Dang, T. N. and Wilkinson, L. (2012). Timeseer: Detecting interesting distributions in multiple time series data. *VINCI*, pages 1–9.

de Jong, J. and Russell, K. (2016). *listviewer: 'htmlwidget' for Interactive Views of R Lists.* R package version 1.2.

Dipert, A., Schloerke, B., and Borges, B. (2018). *shinyloadtest: Load Test Shiny Applications.* R package version 1.0.0.

Dorling D. (1996). Area Cartograms: Their Use and Creation. *Concepts and Techniques in Modern Geography (CATMOG).*

Dougenik et al. (1985). An Algorithm To Construct Continuous Area Cartograms. *The Professional Geographer.*

Emerson, J. W., Green, W. A., Schloerke, B., Crowley, J., Cook, D., Hofmann, H., and Wickham, H. (2013). The generalized pairs plot. *Journal of Computational and Graphical Statistics*, 22(1):79–91.

Few, S. (2006). Data visualization: Rules for encoding values in graph. https://web.archive.org/web/20160404214629/http://www.perceptualedge.com/articles/b-eye/encoding_values_in_graph.pdf.

Freedman, D. and Diaconis, P. (1981). On the histogram as a density estimator: L2 theory. *Zeitschrift für Wahrscheinlichkeitstheorie und verwandte Gebiete*, 57:453–476.

Galili, T. (2016). *heatmaply: Interactive Heat Maps Using 'plotly'.* R package version 0.5.0.

Guha, S., Hafen, R., Rounds, J., Xia, J., Li, J., Xi, B., and Cleveland, W. S. (2012). Large complex data: divide and recombine with rhipe. *The ISI's Journal for the Rapid Dissemination of Statistics Research*, pages 53–67.

Hafen, R. (2016). *trelliscope: Create and Navigate Large Multi-Panel Visual Displays.* R package version 0.9.7.

Hafen, R., Gosink, L., McDermott, J., Rodland, K., Dam, K. K. V., and Cleveland, W. S. (2013). Trelliscope: A system for detailed visualization in the deep analysis of large complex data. In *Large-Scale Data Analysis and Visualization (LDAV), 2013 IEEE Symposium on*, pages 105–112.

Hafen, R. and Schloerke, B. (2018). *trelliscopejs: Create Interactive Trelliscope Displays*. R package version 0.1.3.

Hamilton, N. (2016). *ggtern: An Extension to 'ggplot2', for the Creation of Ternary Diagrams*. R package version 2.1.4.

Harrell Jr, F. E., with contributions from Charles Dupont, and many others. (2019). *Hmisc: Harrell Miscellaneous*. R package version 4.2-0.

Healey, K. (2018). *Data Visualization: A Practical Introduction*. Princeton University Press.

Heer, J., Agrawala, M., and Willett, W. (2008). Generalized selection via interactive query relaxation. In *Proceedings of the SIGCHI Conference on Human Factors in Computing Systems*, pages 959–968. ACM.

Heer, J. and Bostock, M. (2010). Crowdsourcing graphical perception: Using mechanical turk to assess visualization design. In *ACM Human Factors in Computing Systems (CHI)*, pages 203–212.

Hester, J., Müller, K., Ushey, K., Wickham, H., and Chang, W. (2018). *withr: Run Code 'With' Temporarily Modified Global State*. R package version 2.1.2.

Hijmans, R. J. (2019). *raster: Geographic Data Analysis and Modeling*. R package version 2.8-19.

Hyndman, R. J. (2018). *forecast: Forecasting functions for time series and linear models*. R package version 7.2.

Ihaka, R., Murrell, P., Hornik, K., Fisher, J. C., Stauffer, R., Wilke, C. O., McWhite, C. D., and Zeileis, A. (2019). *colorspace: A Toolbox for Manipulating and Assessing Colors and Palettes*. R package version 1.4-0.

Inc, F., Russell, K., and Dipert, A. (2019). *reactR: React Helpers*. R package version 0.3.0.

Jeppson, H., Hofmann, H., and Cook, D. (2016). *ggmosaic: Mosaic Plots in the 'ggplot2' Framework.* R package version 0.0.1.

Jeworutzki, S. (2018). *cartogram: Create Cartograms with R.* R package version 0.1.1.

Karambelkar, B. (2017). *widgetframe: 'Htmlwidgets' in Responsive 'iframes'.* R package version 0.3.1.

Lander, J. P. (2016). *coefplot: Plots Coefficients from Fitted Models.* R package version 1.2.4.

Leeper, T. J. (2017). *slopegraph: Edward Tufte-inspired Slopegraphs.* R package version 0.1.14.

Lins, L., Klosowski, J. T., , and Scheidegger, C. (2013). Nanocubes for real-time exploration of spatiotemporal datasets. *Visualization and Computer Graphics, IEEE Transactions.*

Liu, Z. and Heer, J. (2014). The effects of interactive latency on exploratory visual analysis. *IEEE Trans. Visualization & Comp. Graphics (Proc. InfoVis).*

Liu, Z., Jiang, B., and Heer, J. (2013). immens: Real-time visual querying of big data. *Computer Graphics Forum (Proc. EuroVis), 32(3).*

Lovelace et al. (2019). *Geocomputation with R.* Chapman & Hall/CRC.

Mastny, T. (2018). *sass: Syntactically Awesome StyleSheets (SASS) Compiler.* R package version 0.1.0.9000.

Messing, S. (2012). Visualization series: Insight from cleveland and tufte on plotting numeric data by groups. http://web.archive.org/web/20160602202734/https://solomonmessing.wordpress.com/2012/03/04/visualization-series-insight-from-cleveland-and-tufte-on-plotting-numeric-data-by-groups/.

Mildenberger, T., Rozenholc, Y., and Zasada., D. (2009). *histogram: Construction of regular and irregular histograms with different options for automatic choice of bins.* R package version 0.0-23.

Moritz, D., Howe, B., and Heer, J. (2019). Falcon: Balancing interactive latency and resolution sensitivity for scalable linked visualizations. http://idl.cs.washington.edu/papers/falcon.

Mullen, L. A. and Bratt, J. (2018). USAboundaries: Historical and contemporary boundaries of the united states of america. *Journal of Open Source Software*, 3:314.

Murray, S. (2013). *Interactive Data Visualization for the Web: An Introduction to Designing with D3*. O'Reilly Media.

Murray, S. (2017). *D3.js in Action*. Manning Publications.

Newman, M. (2016). Maps of the 2016 U.S. presidential election results. http://www-personal.umich.edu/~mejn/election/2016/.

Nusrat et al. (2016). Evaluating cartogram effectiveness. *IEEE Trans Vis Comput Graph*, 24:1077–1090.

Olson (1976). Noncontiguous Area Cartograms. *The Professional Geographer*, 28:371–380.

Ooms, J. (2014). The jsonlite package: A practical and consistent mapping between json data and r objects. *arXiv:1403.2805 [stat.CO]*.

Ooms, J. (2018). *rsvg: Render SVG Images into PDF, PNG, PostScript, or Bitmap Arrays*. R package version 1.3.

Ooms, J. (2019). *magick: Advanced Graphics and Image-Processing in R*. R package version 2.0.9000.

Pebesma, E. (2018). *sf: Simple Features for R*. R package version 0.6-0.

Pebesma, E. J. and Bivand, R. S. (2005). Classes and methods for spatial data in R. *R News*, 5(2):9–13.

Pedersen, T. L. (2016). *ggraph: An Implementation of Grammar of Graphics for Graphs and Networks*. R package version 0.1.1.

Pedersen, T. L. (2019). *ggforce: Accelerating 'ggplot2'*. R package version 0.2.0.

PROJ contributors (2018). *PROJ coordinate transformation software library*. Open Source Geospatial Foundation.

Quealy, K. (2013). 19 sketches of quarterback timelines. https://web.archive.org/web/20180416011941/http://kpq.github.io/chartsnthings/2013/09/19-sketches-of-quarterback-timelines.html.

R Core Team (2016). *R: A Language and Environment for Statistical Computing*. R Foundation for Statistical Computing, Vienna, Austria.

Robinson, D. (2016). *broom: Convert Statistical Analysis Objects into Tidy Data Frames*. R package version 0.4.1.

RStudio (2014a). Build custom input objects. https://shiny.rstudio.com/articles/building-inputs.html.

RStudio (2014b). Profvis — interactive visualizations for profiling r code. https://rstudio.github.io/profvis/examples.html.

Rudis, B. (2016). *ggalt: Extra Coordinate Systems, Geoms and Statistical Transformations for 'ggplot2'*. R package version 0.1.1.

Ruiz, E. (2018). *dbplot: Simplifies Plotting Data Inside Databases*. R package version 0.3.0.

Ryan, J. A. (2016). *quantmod: Quantitative Financial Modelling Framework*. R package version 0.4-6.

Sarkar, D. (2008). *Lattice: Multivariate Data Visualization with R*. Springer, New York. ISBN 978-0-387-75968-5.

Schloerke, B., Crowley, J., Cook, D., Briatte, F., Marbach, M., Thoen, E., Elberg, A., and Larmarange, J. (2016). *GGally: Extension to 'ggplot2'*. https://ggobi.github.io/ggally, https://github.com/ggobi/ggally.

Scott, D. W. (1979). On optimal and data-based histograms. *Biometrika*, 66:605–610.

Scott, D. W. (1992). *Multivariate density estimation: theory, practice, and visualization*. John Wiley & Sons.

Shneiderman, B. (1996). The Eyes Have It: A Task by Data Type Taxonomy for Information Visualizations. *VL Proceedings of the IEEE Symposium on Visual Languages*, pages 1–9.

Sievert, C. (2018a). *bcviz: A shiny app for exploring BC Housing and Census Data*. R package version 0.1.

Sievert, C. (2018b). Learning from and improving upon ggplotly conversions. https://blog.cpsievert.me/2018/01/30/learning-improving-ggplotly-geom-sf/.

Sievert, C. (2018c). Visualizing geo-spatial data with sf and plotly. https://blog.cpsievert.me/2018/03/30/visualizing-geo-spatial-data-with-sf-and-plotly/.

Sievert, C. (2018d). *zikar: Tools for exploring publicly available Zika data.* R package version 0.0.0.9000.

Sievert, C. (2019a). *pedestrians: Tools for exploring Melbourne's Pedestrian Data.* R package version 0.0.1.

Sievert, C. (2019b). *runpkg: Tools for working with 'UNPKG'.* R package version 0.0.0.9000.

South, A. (2017). *rnaturalearth: World Map Data from Natural Earth.* R package version 0.1.0.

Sturges, H. A. (1926). The choice of a class interval. *Journal of the American Statistical Association,* 21(153):65–66.

Swayne, D. F., Cook, D., and Buja, A. (1998). XGobi: Interactive Dynamic Data Visualization in the X Window System. *Journal of Computational and Graphical Statistics,* 7(1):113–130.

Theus, M. and Urbanek, S. (2008). *Interactive Graphics for Data Analysis: Principles and Examples.* Chapman & Hall / CRC.

Tidyverse team (2018). Tidyverse design principles. https://principles.tidyverse.org.

Tierney, N., Cook, D., McBain, M., and Fay, C. (2018). *naniar: Data Structures, Summaries, and Visualisations for Missing Data.* R package version 0.4.2.9000.

Tufte, E. (2001). *The Visual Display of Quantitative Information.* Graphics Press, Cheshire, Conn.

Tukey, F. and Fisherkeller (1973). Stanford linear accelerator. http://stat-graphics.org/movies/prim9.html.

Tukey, J. and Tukey P. (1985). Computer graphics and exploratory data analysis: An introduction. In *In Proceedings of the Sixth Annual Conference and Exposition: Computer Graphics85.*

Tyner, S. and Hofmann, H. (2016). *geomnet: Network Visualization in the 'ggplot2' Framework.* R package version 0.1.2.1.

Unwin, A. (2015). *Graphical Data Analysis with R.* CRC Press.

Unwin, A. (2016). GDA of England (from engsoccerdata). http://www.gradaanwr.net/wp-content/uploads/2016/06/dataApr16.pdf.

Unwin, A. and Hofmann, H. (1999). Gui and command-line — conflict or synergy? In Berk, K. and Pourahmadi, M., editors, *Computing Science and Statistics, Proceedings of the 31st Symposium on the Interface,* volume 31, pages 246–253, Chicago. Interface Foundation.

Urbanek, S. (2013). *png: Read and write PNG images.* R package version 0.1m-7.

Urbanek, S. (2015). *base64enc: Tools for base64 encoding.* R package version 0.1-3.

Vaidyanathan, R., Xie, Y., Allaire, J., Cheng, J., and Russell, K. (2016). *htmlwidgets: HTML Widgets for R.* R package version 0.6.

VanderPlas, S. and Hofmann, H. (2015). Signs of the sine illusion—why we need to care. *Journal of Computational and Graphical Statistics,* 24(4):1170–1190.

Venables, W. N. and Ripley, B. D. (2002). *Modern Applied Statistics with S.* Springer, New York, fourth edition. ISBN 0-387-95457-0.

Walker, K. (2018). *idbr: R Interface to the US Census Bureau International Data Base API.* R package version 0.3.

Wickham, H. (2009). *ggplot2: Elegant Graphics for Data Analysis.* Springer-Verlag New York.

Wickham, H. (2010). A layered grammar of graphics. *Journal of Computational and Graphical Statistics,* 19(1):3–28.

Wickham, H. (2013). Bin-summarise-smooth: a framework for visualising large data. Technical report, had.co.nz.

Wickham, H. (2014a). *Advanced R*. Chapman & Hall/CRC.

Wickham, H. (2014b). Tidy data. *The Journal of Statistical Software*, 59.

Wickham, H. (2016). *ggstat: Statistical Computations for Visualisation*. R package version 0.0.0.9000.

Wickham, H. (2018a). *nycflights13: Flights that Departed NYC in 2013*. R package version 1.0.0.

Wickham, H. (2018b). You can't do data science in a gui. https://www.youtube.com/watch?v=cpbtcsGE0OA.

Wickham, H., Cook, D., and Hofmann, H. (2015). Visualizing statistical models: Removing the blindfold. *Statistical Analysis and Data Mining: The ASA Data Science Journal*, 8(4):203–225.

Wickham, H., François, R., and D'Agostino McGowan, L. (2018). *emo: Easily Insert 'Emoji'*. R package version 0.0.0.9000.

Wickham, H. and Grolemund, G. (2016). *R for Data Science*. O'Reilly Media.

Wilke, C. (2018). *Fundamentals of Data Visualization*. O'Reilly Media.

Wilkinson, L. (2005). *The Grammar of Graphics (Statistics and Computing)*. Springer-Verlag New York, Inc., Secaucus, NJ, USA.

Wilkinson, L., Anand, A., and Grossman, R. (2005). Graph-theoretic scagnostics. In *Proceedings of the Proceedings of the 2005 IEEE Symposium on Information Visualization*, INFOVIS '05, pages 21–, Washington, DC, USA. IEEE Computer Society.

Wilkinson, L. and Wills, G. (2008). Scagnostics distributions. *Journal of Computational and Graphical Statistics*, 17(2):473–491.

Wills, G. (2008). *Linked Data Views*, pages 217–241. Springer, Berlin, Heidelberg.

Xie, Y. (2016). *servr: A Simple HTTP Server to Serve Static Files or Dynamic Documents*. R package version 0.4.

Xie, Y. (2018). Using DT with shiny. https://rstudio.github.io/DT/shiny.html.

Yau, N. (2011). *Visualize This: The FlowingData Guide to Design, Visualization, and Statistics.* John Wiley & Sons.

Yau, N. (2016). What i use to visualize data. `https://flowingdata.com/2016/03/08/what-i-use-to-visualize-data/`.

Index

9 781138 331495